CW01668993

Means of Escape from Fire in Buildings

Means of escape from fire in buildings

A Design Guide to the Statutory Requirements (England and Wales)

J I'A Nelson

CHARLES KNIGHT PUBLISHING
CROYDON

Published by
Charles Knight Publishing, Tolley House, 2 Addiscombe Road, Croydon CR9 5AF

A United Newspapers Publication

Second (revised) impression November 1990

Printed in Great Britain by
Dotesios Printers Ltd,
Trowbridge, Wilts

ISBN 0 85314 372–2

Preface

Statutory requirements for means of escape from buildings in the event of fire arise from a number of Acts of Parliament or regulations made under an Act. The aim of all of them is ensuring the safety of the people who may be legitimately occupying the building as residents, employees, workers, for entertainment or for any other purpose. It is not the function of Statutory requirements to protect property (the building or its contents) to a greater extent than may be necessary to ensure the safety of people in or around a building.

Chapter 1 of this design guide reviews the principal Statutes which require adequate means of escape to be provided from buildings in England and Wales. The principles upon which the design of means of escape should be based are reviewed, introducing the reader to terms and performance standards. Chapter 1 also indicates what guidance is available for those who, as designers of buildings, must comply with the legislation, or those who have the duty to approve designs prepared by others.

Chapter 2 reprints those sections of Knight's Building Regulations which explain the means of escape requirements of building regulations made under the Building Act 1984. These requirements are in the form of mandatory rules, stating a functional requirement and quoting certain clauses in British Standard Codes of Practice as the means of satisfying the functional requirement. The clauses relate directly to public safety within the scope of the Building Act. Complying with Code recommendations is a Statutory requirement of the Building Regulations.

The scope of the Fire Precautions Act 1971, as subsequently amended, is wider than the scope of the Building Act 1984. All premises put to a designated use must be provided with satisfactory means of escape and of fighting fire; in the majority of cases the owner or occupier is required to have a Certificate covering these and certain other matters. Some premises or types of use may be exempted either by the Act or by the fire authority from the need to be certified.

Conditions attached to the granting of a fire certificate include subjects not dealt with in Building Regulations (for example, provision of a fire alarm system). No regulations comparable with Building Regulations have been made under the Fire Precautions Act but the latest amendments empower the Secretary of State to issue Codes of Practice which will have a status similar to an Approved Document. Otherwise requirements are largely at the discretion of the fire authority issuing the certificate.

A number of authoritative guidance documents have been issued for different classes of building and these are listed in Chapter 1. It will be found that the principles on which BS Codes, departmental codes and other advisory documents are based are the same; these principles are set out briefly in Chapter 1.

When designing satisfactory means of escape from buildings, special consideration must be given to the needs of the disabled. BS 5588: Part 8 is the Code of Practice which deals with this subject; the recommendations are summarised in Chapter 3. It should be noted that BS 5588: Part 8 is not intended to be applied to dwellings and flats and satisfying the requirements is not made mandatory by Building Regulations.

There are important sections in each of the British Standards which, because they are not cited in the Mandatory Rules, are not reviewed in Chapter 2. Some sections deal with the provision of fire alarms and means of fighting fire and are therefore relevant to the Fire Precautions Act requirements. Other sections deal with matters beyond the scope of present legislation; they are nonetheless important. They include, for example, advice to building managers about the maintenance of fire precaution facilities and the training of staff in fire drill.

Finally I would like to acknowledge the assistance I have received in writing this book from the helpful comments made by Eric Thompson and Harold Clarke, my co-authors of *Knight's Building Regulations*, particularly in the way Harold has kept me up to date with changes in the legislation. I also appreciate the efforts the staff at Charles Knight have made in preparing the book for publication.

J. I'A. Nelson,
Rickmansworth,
September 1989

Contents

CHAPTER 1

General Statutory Requirements

1 MEANS OF ESCAPE

The term "means of escape" is normally defined as "the structural means whereby persons can escape in the event of fire from any part of a building to a place of safety without outside assistance". When applied to new works, the word "structural" means features which are an integral part of the construction of the building. Items such as rope escapes, chutes or similar "add-on" devices have no part in structural means of escape. There may be circumstances, however, when improving an existing building when satisfactory means of escape can only be achieved by the use of non-structural devices or external additions to the building.

The phrase "without outside assistance" means that occupants should be able to leave the building without having to be rescued by the fire service; for example, by ladder. Inside assistance may be needed, however, for disabled persons or those confined to a hospital bed; this assistance should be provided by trained, able-bodied staff nominated by management.

The "place of safety" should be a place away from the building in the open air at ground level. Only exceptionally (for example, in a multi-occupancy complex) will a place within another building, in a common access mall or in an enclosed space be considered a place of safety. A recent amendment to the Fire Precautions Act 1971, contained in the Fire Safety and Safety of Places of Sport Act 1987, has defined "escape" so as to allow a fire certificate to impose conditions which apply beyond the curtilage of the building. This is to ensure that occupants escaping through the final exit can move onwards freely until they are well clear of any risk from the building fire.

Building Regulations apply to the structure of a building but not normally to any part of a site around a building. However, mandatory rules make certain clauses of the four British Standard Codes of Practice for means of escape mandatory. In the Codes the definition of "final exit" and the advice given about site planning ensure that the external part of the escape route must be planned to allow occupants to move to a place where they are no longer at risk.

2 LEGISLATION

Much of the legislation relating to means of escape from buildings in case of fire will be found in two Acts of Parliament, the Fire Precautions Act 1971 and the Building Act 1984. Other Acts make provision for escape in more general terms from classes of building not covered by the main Acts. They have effectively the same result. In some areas additional precautions may be required under a Local Act particulars of which may be obtained from the local authority.

2.1 The Fire Precautions Act 1971

The Fire Precautions Act 1971 has been amended or modified by other Acts and Orders over the years, the latest being the Fire Safety and Safety of Places of Sport Act 1987. The purpose of the 1971 Act, however, remains the same; it is "an Act to make further provisions for the protection of persons from fire risks; and for purposes connected therewith".

The Act applies only to classes of building which have been "designated" by Order. Order 1989 No 76 was the latest to designate factories, shops, offices and railway premises; Article 2 of the Order defined the four uses and other articles gave details of exemptions and dealt with the case of a building divided into two or more parts, each part being in separate ownership or occupation.

The owner or occupier of a designated building, which is not exempted by the Order, must apply to the fire authority for a fire certificate. In some circumstances the fire authority may decide that a fire certificate is not necessary; they have the power to exempt such a building. Otherwise, the authority will inspect the premises and, if satisfied, will issue a fire certificate. Order 1989 No 77 reproduces a model form on which to apply for a fire certificate. If an inspection reveals unsatisfactory features the inspector will state what measures are needed to bring the building up to the required standard. This must be done within a specified time. If at the end of that time (or such further time as may be allowed) a certificate has not been issued, it shall be deemed to have been refused.

An occupier is under a duty, pending the disposal of his application, to ensure that –

(a) the means of escape in case of fire with which the premises are provided can be safely and effectively used at all times;
(b) the means for fighting fire on the premises are maintained in efficient working order; and
(c) any persons employed to work in the premises receive instruction or training in what to do in case of fire.

If, when it was erected, the building was subject to Building Regulation control and plans for it were deposited with the local authority; and also, provided the regulations for that class of building included structural fire requirements and the provision of means of escape, the fire authority may not insist on structural changes as a condition of the granting of a certificate. The inspector would need to have strong reasons for demanding standards higher than are recommended in the Approved Document B 'Fire Spread' or recommended by a BS Code of Practice invoked by the Mandatory Rules. This bar does not apply to matters outside the scope of Building Regulations; e.g. the fitting of fire alarms, fire-fighting equipment, or emergency lighting.

A fire certificate will identify the building and specify the purpose for which it may be used; it will state the maximum number of occupants allowed to use the building. It will indicate, usually on plans, the approved means of escape, the location of fire doors and how the routes should be maintained to keep them safe and usable at all times. The certificate may also specify such features as emergency lighting and the type and placing of direction signs. The type of fire-fighting equipment and of alarm systems to be installed may be specified. The certificate will also deal with the maintenance of escape routes and equipment so that all will be available and usable in an emergency. The training of personnel in fire drills is another matter which may be dealt with in the certificate.

Factories, offices, shops and railway premises

The Fire Precautions (Factories, Offices, Shops and Railway Premises) Order 1989 (SI 1989 No 76) designates buildings which are wholly, or in part, put to one of the above uses. The Order generally exempts from the requirement to have a certificate premises occupied for the above purposes in which not more than 20 persons in total will be employed to work and in which not more than 10 persons will be employed on a floor other than the ground floor. This exemption does not apply when two or more designated premises exist in the same building and the aggregate number of employees exceeds either of the above two limits. Nor does the exemption apply if explosives or highly flammable materials are stored or used and the fire authority consider they constitute an additional risk to persons on the premises. In this case special premises regulations enforced by the Health and Safety Executive are likely to apply. (See SI 1976 No 2003.)

Order No 76 also specifies further categories of building which the fire authority may, if they think fit, exempt from the certification requirement. This power to exempt was inserted into the Fire Precautions Act 1971 as s.5A by Article 1(4) of the Fire Safety and Safety of Places of Sport Act 1987.

In the case of factories, shops and railway premises the fire authority's power to exempt may be applied when the premises are confined to a ground floor, a ground and first floor, ground and basement, or ground, basement and first floors. In the case of shops the power to exempt may be applied when there is a ground floor only, or when there is a ground floor and basement with half-hour fire-resisting construction separating them.

In determining whether the means of escape is adequate, the fire authority will take into consideration the use of the building, the maximum number of people working in it and their location in the building. The inclusion of reliable fire detection, fire alarm and fire-fighting systems will influence the inspector's judgement. Note that "persons at work" in premises include both directly employed and self employed persons as well as those in training, including apprentices.

Exempted premises may have the exemption withdrawn (after suitable notice has been given) by the fire authority. This might be done because the number of employees increased, the process or materials used changed or the premises were subject to alteration or extension. Before any of these events occur, it is the duty of the occupier to notify the fire authority that change is about to take place.

Even if premises are exempt under the Act, or have been granted exemption by the fire authority, the Act imposes a duty on the occupier to ensure that reasonable provision is made for means of escape and means of fighting fire. There are penalties for contravening this duty.

Since Order 1987 No 1762 the fire authority has been empowered to charge a fee for the work involved in issuing (and in some cases amending) a fire certificate; the amount of the fee will be related to the cost of the work.

HOTELS AND BOARDING HOUSES

By the Fire Precautions (Hotels and Boarding Houses) Order 1972 (SI 1972 No 238) a fire certificate is required for all hotels and boarding houses in which sleeping accommodation is provided on the premises for more than 6 persons (staff or guests). A certificate is also required if any sleeping accommodation is provided in such premises on or above the second storey, or in a basement.

The procedure for applying for a certificate and the conditions to be satisfied to obtain it are similar to those applying to factories, etc.

2.2 The Building Act 1984

BUILDING REGULATIONS

Section 1 of the Building Act 1984 gave the Secretary of State for the Environment the power to make Building Regulations. Paragraph 7(a)(iv)(c) of Schedule 1 of the Act permits the making of regulations which provide for "means of escape in case of fire . . .". Such regulations came into effect in 1985. The provisions for means of escape from fire are to be found in Part B1 of the Schedule of technical requirements which are expressed in functional terms. B1 states that "there shall be means of escape in case of fire from the building to a place of safety outside capable of being safely and effectively used at all material times". B1 also states that the requirement can only be satisfied by complying with a set of mandatory rules. The rules are issued under separate cover.

The Schedule applies the functional requirement to the following building occupancies –

- A building which is, or contains, a DWELLINGHOUSE of three or more storeys. An Appendix to the Mandatory Rules makes special requirements which apply when a two-storey dwellinghouse has the roof space converted to form one or two habitable rooms.
- A building which contains a FLAT and is of three or more storeys.

In these two cases the means of escape need only be provided for the occupants on the third storey and any storey above it.

- A building which is, or contains, an OFFICE.
- A building which is, or contains, a SHOP.

In the case of an office or a shop, the means of escape need only be provided for the occupants of the office or shop as the case may be.

The Mandatory Rules state that the requirement in B1(1) can only be met by complying with the recommendations of certain clauses of appropriate British Standards. Appendix B of the Rules contains specific requirements for roofspace conversions in two-storey dwellinghouses; these represent a relaxation of the recommendations for three-storey dwellings contained in the BS Code. The relevant requirements of the four British Standard Codes of Practice* and the requirements of the Appendix are summarised in Chapter 2.

OTHER SECTIONS OF THE BUILDING ACT

Other sections of the Act make requirements for means of escape from buildings; they apply to buildings which are not already subject to the requirements for B1 or the fire certificate provisions, but in which there is a risk due to the assembly of large numbers of people or a sleeping risk. The sections are ss.24 and 71 (the provision of exits from certain buildings) and s.72; they impose a duty on local authorities to ensure that the means of escape is adequate.

Sections 24 and 71 are concerned with "means of ingress and egress and passages or gangways". Section 24 applies when full plans of a building or extension are deposited with the local authority; section 71 applies to a building, whether or not full plans were deposited before it was built. In both cases the means of ingress, etc. must be "deemed satisfactory".

The sections require the local authority to consult the fire authority; this implies that ingress, egress and passageways must, among other things, satisfy fire safety standards. Section 24 lists the occupancies to which both sections apply. In carrying out their duty under this section the local authority will take fire authority advice about whether the building in question has satisfactory egress, etc., having regard to its purpose and the number of persons likely to be in it at any time. If not, the authority shall, by notice, require such work to be done as will put matters right.

The list of occupancies to which these two sections are applied is as follows –

- A theatre, hall or other building used as a place of public resort.
- A restaurant, shop, store or warehouse to which members of the public are admitted and in which more than 20 people are employed.
- A club which is required to be registered under the Licensing Act 1964 (premises in which intoxicating liquor is sold for consumption on the premises and registered clubs).
- A school which is not exempted from building regulations: (exempted schools are those subject to the Education Act 1980; *vide* s.4(1) of the Building Act 1984).
- A church, chapel or other place of public worship, but not:
  - a private house to which members of the public are admitted occasionally or exceptionally;
  - a building already used as a church or chapel before s.36 of the Public Health Acts Amendment Act 1890, (or a corresponding provision of a local Act), came into operation in the district;
  - a building so used before 1st October 1937 in a district where neither of the above has applied.

Section 72 of the Building Act 1984 requires a local authority (also in consultation with the fire authority) to ensure that adequate means of escape from fire is available in certain other buildings, actual or proposed. They are buildings in which people sleep and which exceed two storeys in height. Means of escape must be provided from any floor the level of which is 20 ft (just over 6 m) above the level of the pavement or ground on any side of the building. The buildings are –

- A building let in flats or tenements.

---

* See comment about Codes on page 9.

- A building used as an inn, hotel, boarding house, hospital, nursing home, boarding school, children's home, or similar institution.
- A building used as a restaurant, shop, store or warehouse which has sleeping accommodation for employees on an upper floor.

This section, like s.71, does not apply to buildings for which the Building Regulations impose means of escape requirements, nor where there is a fire certificate in force.

RELAXING OR DISPENSING WITH REQUIREMENTS

The Building Act 1984 provides for requirements which are unreasonable or too onerous in particular circumstances to be relaxed or dispensed with altogether. Section 8 of the Act enables this to be done and Regulation 10 delegates the power to do so to the local authority. As most Building Regulations are expressed in functional terms, the relaxation procedure is rarely needed. However, means of escape requirements derived from Building Regulations are specific; Mandatory Rules prescribe compliance with the advice of BS Codes of Practice and the Codes impose many specific requirements. However, the introduction to the Mandatory Rules states ". . . if in any particular instance the Requirements seem to be particularly demanding, the local authority may be asked to relax them."

There may be circumstances when it would be sensible to dispense with a requirement and this power is also delegated to the local authority. In the event of any dispute concerning relaxation or dispensation, there is a right of appeal to the Secretary of State.

In the event of a dispute between a local authority and a person proposing to carry out work as to whether plans are in conformity with building regulations, application for a determination may be made to the Secretary of State under s.16(10) of the 1984 Act.

2.3 Other Acts

In addition to the two major Acts described above, there are Statutory Instruments made under other Acts which require means of escape to be provided. They apply to certain existing buildings used (or intended to be used) for particular purposes. The principal ones are as follows.

HOUSES IN MULTIPLE OCCUPATION

The "Housing (Means of Escape from Fire in Houses in Multiple Occupation) Order 1981 (S.I. 1981 No 1576)" made under the Housing Act 1980 applies to houses of three or more storeys (excluding any basement) in which the aggregate area of all floors (including the area of stairways and basement storeys) exceeds 500m^2. The Order provides that a local authority may, and in prescribed circumstances must, require the execution of works to provide adequate means of escape for the occupants. In complying with this Order, the local authority will consult the fire authority, who will advise on how the means of escape can be made satisfactory.

LICENSED PREMISES

The provision of adequate means of escape from fire may be made a condition of the grant of a licence for premises which are subject to Licensing or Registration provisions. These include such occupancies as betting offices and public houses on the one hand and nursing homes and residential homes for the elderly or the disabled on the other.

HOSPITALS AND RESIDENTIAL CARE HOMES

National Health Service hospitals are Crown Buildings and not subject to either the Building Act or Fire Precautions Act. However, the design of NHS hospitals must follow guidance issued by the Home Office. Fire precautions, including arrangements for means of escape, are dealt with by HM Inspector of Fire Services at the Home Office. The responsibility for inspecting a completed hospital to ensure that requirements have been met will usually be delegated to the local fire authority for the district in which the new hospital is located.

Private hospitals and nursing homes have not been designated under the Fire Precautions Act 1971 but they do have to be registered under the Registered Homes Act 1984 and they are subject to the Nursing Homes and Mental Nursing Homes Regulations 1984 (SI 1984 No 1578 as amended by SI 1986 No 456). These regulations require the person registered to take adequate fire precautions including the provision of facilities for the evacuation of patients and staff in the event of fire.

The Residential Care Homes Regulations (SI 1984 No 1345 as amended by SI 1986 No 457) also require that there be consultation with the fire authority concerning the provision of adequate fire precautions, including means of escape.

Schools

Part V (Regs 22 to 27) of the Education (School Premises) Regulations (SI 1981, No 909) includes requirements for the provision of means of escape from fire in day schools maintained by local authorities.

Sports grounds

The Safety of Sports Grounds Act 1975, as amended by the Fire Safety and Safety of Places of Sport Act 1987 requires designated sports grounds with accommodation for more than 10,000 spectators to have a safety certificate. In this case it is the local rather than the fire authority to whom application must be made and who will issue the certificate. Designation Orders have been made naming specific football and rugby grounds and certain cricket grounds.

The certificate must state the number, size and location of exits and details of arrangements for means of escape. All grounds subject to certification must be inspected by the local authority at least once a year.

For sports grounds which are not designated, Part III of the Fire Safety and Safety of Places of Sport Act 1987 provides for a safety certificate to be obtained from the local authority for any stand providing covered accommodation for more than 500 spectators. Although this Part of the Act does not specifically mention means of escape from a stand in the event of fire, it is an aspect of "safety" likely to be considered by the local authority inspectors.

3 ADVICE ON MEANS OF ESCAPE

Whichever piece of legislation applies and whether it is the local or the fire authority who will administer it, requirements will either be based on the recommendations of a BS Code of Practice, if one exists for the particular class of building, or on other advisory documents (see below). At present Codes are available for dwellings of three or more storeys, buildings containing flats and for offices and shops. The recommendations of each of those Codes are reviewed in Chapter 2.

The Codes are intended to provide advice to the designer of new buildings; when existing buildings are being altered or converted, it may not be possible to follow all the recommendations of the appropriate Code. In such cases, it is essential to obtain professional advice; to adopt only such advice in a Code as suits the work being undertaken, while ignoring the rest, is likely to give an unsatisfactory, if not dangerous, result.

For some classes of building not the subject of a BS Code, guidance is available from HMSO. The following documents will be found to contain useful advice –

Education buildings:	DES Building Bulletin No 7, "Fire in the design of schools", HMSO.

Hotels and boarding houses:	Home Office Guide to the Fire Precautions Act 1971, No 1, HMSO.
Factories:	Guide No 2, Part II in the Home Office series, HMSO.
Offices, shops and railway premises:	Guide No 3, Part II in the Home Office series, HMSO.
Hospitals:	Hospital design note No 2, "Protection against fire", HMSO.
Town centre redevelopment:	Home Office Fire Prevention Guide No 1, "Fire precautions in town centre redevelopment", HMSO.

4 PRINCIPLES OF MEANS OF ESCAPE FROM FIRE

4.1 Objectives and standards

Designing means of escape means providing, within the building, escape routes in sufficient number, located so as to serve all occupied areas of the building and wide enough to cope with the numbers of people likely to need to use them in an emergency. Wherever possible occupants should have alternative means of escape available which allow escape to be made in significantly different directions; by this means a person escaping may move away from the source of fire. As much of the route as possible should be protected from the ingress of fire and smoke by enclosing it in fire-resisting structure. Designing means of escape should be undertaken as part of the wider problem of planning the circulation within the building.

The means of escape should permit all occupants threatened by fire to leave the building (or compartment) on fire immediately, in an orderly manner, without the need for outside assistance and to reach a place of safety. The place of safety may be another compartment but is more usually in the open air at ground level. It may be necessary for everyone in the building to leave it, or only those in the compartment where fire has occurred. It is usual to assume that all occupants at risk should be able to reach a place of safety in 2.5 minutes.

Codes of Practice and design guides interpret these aims by making recommendations for planning and layout and by setting standards for a number of key factors. The factors are interrelated and the importance of one relative to the others varies according to the type of occupancy, building layout, and the number, distribution and state of mind and body of the occupants. They are as follows –

- *Direct distance and travel distance:* These are the maximum distances, measured either in a straight line ignoring obstructions, or along the actual route an occupant would have to travel to get from a point on a storey to a doorway leading to relative safety, or to a final exit from the building. The permitted maximum distance will be greater if two well separated, alternative routes are available (alternative means of escape). Codes may also recommend that doors be located at intervals along a corridor to restrict the spread of smoke.
- *Minimum number of exits:* Codes will recommend how many exits there should be from a dwelling, or a part of a storey, or an entire storey. The number required is dependant on planning, or number of occupants, or travel limitations or acceptable exit widths. It is usual to assume that one exit from a storey may be unusable because of the fire; the number is also influenced by the need to provide alternative means of escape. The number needed to serve a storey in a multi-storey building will dictate the number of stairways needed. There are some circumstances in certain occupancies, when it may be possible to accept a single stairway.
- *Location of exits:* Exits should be located so that travel distance limits may be adhered to and alternative means of escape provided. Final exits should be

located away from high risk areas such as plant rooms and where those escaping can move quickly away from the building to a place of safety.

- *Width of exits and escape routes:* Doorways, corridors and stairways must be wide enough for the flow of people needing to use them. 'Width' may be of an individual exit or the aggregate width of all exits serving a storey; width of stairways will also be specified. Studies of the flow of people through doors, corridors and on stairways has enabled width to be related to number of users. Important factors which affect stairway width are how many storeys of the building will need to be evacuated, the number that can be accommodated within the stairway enclosure between storeys and the need for intermediate handrailing if the width exceeds a maximum figure.
- *Number of occupants:* The number and aggregate width of exits depend on how many occupants there are on each storey. In areas with fixed seating or known capacity such as auditoria or restaurants, design may be based on actual numbers. Elsewhere the Codes contain advice on "occupancy factors" – number per m^2 of floor area; this varies according to the class of building and the use to be made of an area or room.

4.2 Designing means of escape

To achieve satisfactory means of escape from fire, the designer should first consider the nature of the fire hazards. The fire hazard in a building occupancy arises from a number of contributing factors; each factor has differing significance depending on the type of building (construction, height, spacial design, etc.), its function, services and equipment contained in it and the nature of the occupants. The function of the building, especially the sort of activities and operations carried on in it and materials being processed, may constitute a hazard.

The hazard due to occupants can vary considerably. A large number of people in a confined space, as in assembly buildings, constitutes a greater hazard than do few people dispersed in a large volume. The kind of occupant, whether physically or mentally handicapped and whether familiar or not with the layout of the building, is significant, as is the "sleeping risk" in a residential occupancy.

The design and construction of the building increase or reduce hazard according to: the materials used, especially for exposed finishes; internal planning, particularly if large open spaces and interconnected storeys are incorporated; the size of compartments and the height of the topmost storey above ground level.

Some hazards can be offset by designing active fire precaution measures into the building. Those which significantly reduce the hazard to occupants include an approved fire detection system, an alarm system and built-in fire fighting equipment, either manual or automatic.

All the Codes recommend that a designer should consider how a fire, starting at any point in the building, might grow and spread. This involves identifying potential sources, assessing likely fire severity and forecasting probable development. Statistical information is available about the incidence and sources of fire in different occupancies, but it can generally be assumed that a fire can start almost anywhere in the building. It can be assumed that it is unlikely to start at two locations at the same time.

It is important to consider how a fire might spread. This enables fire protective features such as compartment walls, floors and fire doors to be incorporated to contain fire and restrict spread. Materials of construction can be selected to minimise the contribution they could make to the fire. Identifying areas of higher than normal risk, isolating them and enclosing them reduces hazard. Means of escape is then located in the safer parts of the building.

A major principle in the design of means of escape is the provision of alternative escape routes, starting in significantly different directions from most locations in the building. Exceptions to this principle are few and additional precautions must generally be taken. In a small room, with few occupants, a single exit will suffice, but once outside the room alternatives should be on offer. The

Codes also specify closely the circumstances which make "dead-end" areas of a storey acceptable.

Another aspect of design concerns the exclusion or removal of smoke from the route. The primary danger to occupants during the escape phase is not heat or flame but smoke and toxic gases. If an internal corridor forms part of an escape route, fire doors across the corridor may be used to restrict smoke spread. Alternatively, the corridor may be ventilated at each end (and at intermediate positions, if necessary) to remove smoke. It may be necessary to make a length of corridor a protected zone by using fire resisting construction for the walls, floors and doors enclosing it.

Stairways forming part of an escape route are always constructed as fully protected zones and it is vital to exclude smoke from them. Fire doors at every storey level and provision for ventilation are usually sufficient, but in some circumstances a ventilated approach lobby may be required. Pressurisation may also be used to exclude smoke from stairways.

5 THE BRITISH STANDARD CODES OF PRACTICE

Chapter 2 describes the recommendations of four British Standards, certain clauses of which become Building Regulation requirements by virtue of the Mandatory Rules. The relevant clauses are listed in an appendix at the end of each section of Chapter 2. The British Standards are –

BS 5588: Section 1.1: 1984. Code of practice for single-family dwellinghouses
BS CP 3: Chapter IV: Pt 1: 1971. Flats and maisonettes (in blocks over two storeys)
BS 5588: Pt 2: 1985. Code of practice for shops
BS 5588: Pt 3: 1983. Code of practice for office buildings

Two of these Standards, Section 1.1: 1984 and CP 3: Chapter IV, have been replaced since the publication of the Mandatory Rules by BS 5588: Part 1: 1990 "Code of Practice for Residential Buildings". It would be sensible to consult this latest version for advice about fire precautions in the design of dwellings, but, until the relevant part of Schedule 1 of the 1985 Building Regulations – clause B1(2) – is amended, it is the old versions which have the force of law under the Mandatory Rules.

In addition to the four Codes listed above, BS 5588: Pt 8: 1988 is relevant to the design of means of escape. It is the Code of practice for means of escape for disabled people. The Code recommendations are reviewed in Chapter 3.

The purpose of both the Building Act and the Fire Precautions Act is to make provision for the safety of people in and around buildings. None of the legislation is aimed at protecting the building or its contents more than is needed to ensure personal safety. The British Standards have a broader scope and thus contain many clauses which are not cited in the Mandatory Rules. Only the clauses cited in the Mandatory Rules are reviewed in Chapter 2.

The non-statutory sections of the Codes give advice to building owners and recommend certain fire protection facilities. The advice to owners will, it is hoped, reduce the likelihood of fire breaking out; it encourages better "housekeeping" and advises on the institution of fire drills and the training of staff to expedite escape in the event of fire. The section dealing with fire protection facilities includes advice about fire alarms and fire-fighting equipment.

Compliance with both means of escape and constructional requirements prescribed under the various Acts described here, and the inclusion in a building design of the fire protection facilities recommended in the Codes will not only protect persons from fire but will also give a large measure of protection to the building and its contents.

CHAPTER 2

BUILDING REGULATIONS

Mandatory Rules and Codes of Practice

INTRODUCTION

Means of Escape requirements derive from B1 of Schedule 1 to the Building Regulations; they differ from other requirements of the Building Regulations because they can only be satisfied by complying with Mandatory Rules. The Rules are made mandatory by virtue of clause B1(2) of the Schedule. However, in the introductory paragraphs on page 1 of the Rules, it is stated that ". . . if in any particular instance the Requirements seem to be particularly demanding, the local authority may be asked to relax them."

The purpose of the Rules for Means of Escape is to ensure that, as far as is possible by structural means, all occupants of a building will be able to reach a place of safety in the event of fire. The Rules are set out in four sections which correspond to four classes of occupancy. Buildings not represented by these sections may be the subject of other legislation requiring the provision of means of escape. The four classes of occupancy are:

Section 1 DWELLINGHOUSES of three or more storeys
These must comply with relevant clauses of BS 5588: Section 1.1: 1984. An Appendix to the Mandatory Rules makes special provisions for loft extensions to two-storey dwellinghouses.

Section 2 FLATS in buildings of three or more storeys
These must comply with relevant clauses of CP 3: Chapter IV: Pt 1: 1971.

Section 3 SHOPS
These must comply with relevant clauses of BS 5588: Pt 2: 1985.

Section 4 OFFICES
These must comply with relevant clauses of BS 5588: Pt 3: 1983.

The British Standard documents referred to above are Codes of Practice. Throughout these notes, in each of the four Sections, the relevant British Standard for each Section will be referred to as "the Code". The relevant clauses of the Codes are listed in Appendices to the four Sections of the guidance notes which follow.

Two of these Standards, Section 1.1: 1984 and CP 3: Chapter IV, have been replaced since the publication of the Mandatory Rules by BS 5588: Part 1: 1990 "Code of Practice for Residential Buildings". It would be sensible to consult this latest version for advice about fire precautions in the design of dwellings, but, until the relevant part of Schedule 1 of the 1985 Building Regulations – clause B1(2) – is amended, it is the old versions which have the force of law under the Mandatory Rules.

An Appendix to the Rules defines terms used: some terms are also defined in each of the Codes but the definitions do not always agree. It must be assumed that a definition in the Mandatory Rules takes precedence over any which may be given for the same term in any of the Standards.

Application of Regulation B1 and the Mandatory Rules

General limits on application of B1 are given in italics in the Schedule of requirements and are repeated on page 1 of the Mandatory Rules. Specific application information is given in the first paragraph of each of the four Sections. The effect of all these paragraphs is summarised in Table 1 (page 11).

Note that para. 2 of the general "Limits on Application" section attached to Requirement B1 states that "the means of escape provided need only, in the case

Table 1 The application of requirements of B1 of Schedule 1

<table>
<tr><td rowspan="2"></td><td rowspan="2"></td><td>BUILDING USE</td><td>MEANS OF ESCAPE</td><td colspan="3">The Mandatory Rules apply to a building which–</td></tr>
<tr><td>The building is (or contains) –
(1)</td><td>To be provided for the occupants of –
(2)</td><td>is newly erected
(3)</td><td>is extended or materially altered
(4)</td><td>Has its use changed to that in col (1)
(5)</td></tr>
<tr><td rowspan="2">1</td><td>(a)</td><td>a dwellinghouse with 3 or more storeys</td><td>the third and higher storeys</td><td>yes</td><td>yes</td><td>yes</td></tr>
<tr><td>(b)</td><td>a two-storey dwellinghouse extended to provide one or two habitable rooms in the roof</td><td>the habitable rooms in the roofspace</td><td>–</td><td>yes</td><td>–</td></tr>
<tr><td>2</td><td></td><td>of three or more storeys and contains a flat</td><td>any flat on the third or higher storey</td><td>yes</td><td colspan="2">Not the mandatory rules. Section 72 of the Building Act 1984 may apply if the building is let in flats or tenement dwellings.</td></tr>
<tr><td>3</td><td></td><td>a shop</td><td>the shop</td><td>yes</td><td colspan="2" rowspan="2">Not the mandatory rules. The Fire Precautions Act 1971 applies in these cases. Full plans must be deposited to enable the local authority to comply with s. 16 of the F.P.A.</td></tr>
<tr><td>4</td><td></td><td>an office</td><td>the office</td><td>yes</td></tr>
</table>

of a dwellinghouse or a building containing a flat, afford escape from the third storey and above . . ." In the case of a building containing an office or a shop, the means of escape is to afford escape for people from the office or shop.

Note also that it is only in the Section dealing with Dwellinghouses that reference is made to an existing building being extended or materially altered. Regulation 3(2) defines "material alteration" as work which would, at any stage, affect the extent to which the existing building complies with Parts A, B1, B3, B4 and M of Schedule 1 to the Regulations. Thus, building work carried out to extend or alter an existing building must not adversely affect it in respect of compliance with those named Parts (relevant requirements).

The existing building may have been built under former regulations or Byelaws and may comply wholly, partly or not at all with current relevant requirements. To establish whether such an existing building will be adversely affected by proposals, regulation 2(4) requires a comparison to be made of the extent to which the existing building, were it a proposal under current regulations, would satisfy the relevant requirements with the extent to which the proposed building, as altered or extended, satisfies those requirements.

The Department of the Environment has expressed the opinion in a memorandum dated January 1986 sent to some local authorities that the effect of the paragraph 1(a) of the "Limits on application" in B1 is to apply the requirement not only to a building in classes (i) to (iv) which is erected but also to one which is extended or materially altered.

It is considered that this wider interpretation, which has no legislative effect, cannot be supported for the following reasons:

(1) If the words applying the rules to "a building which is erected" were to be read as including buildings to be extended or materially altered, it would not be necessary to include paragraph 1(b) with its specific reference to the extension and material alteration of a dwellinghouse. The fact that 1(b) has been included would indicate that the words "a building which is erected" are intended in the rules to have their ordinary literal meaning. The DoE memorandum suggestion that the meaning is linked to the definition of "building work" in regulation 3 cannot be sustained.

(2) Other differences in the wording of the application paragraphs in the remaining three Sections of the Rules also suggest that the rules are not intended to be applied to extensions and alterations of existing buildings containing a flat on the third storey or above, or containing a shop or an office. In paragraph 1.1 of section 1 the phrase "being erected" infers new construction for the reasons given above. Section (2) only applies to "a building of three or more storeys which is being erected" and would therefore also appear to be limited to new construction. In Sections 3 and 4, paragraphs 3.1 and 4.1 apply the Sections to a building which is or contains . . . a shop . . . an office, but paragraphs 3.2 and 4.2 use the words "is to be constructed" which also imply the erection of a new building.

This interpretation can be further justified because many of the recommendations of the relevant Codes would be either impracticable or very costly to incorporate into many existing buildings.

Changes made by way of extensions or material alterations to buildings which are put, or are to be put, to a designated use will invoke the provisions of the Fire Precautions Act 1971 (as amended). They will be subject to inspection and the fire certification procedure and this will ensure that adequate means of escape, among other things, are provided.

Other considerations

When any building work is proposed, including alterations, extensions or change of use, in respect of which requirement B1 applies, a building notice cannot be given and full plans must be deposited with the local authority for approval (BR.11(2)).

Appendix B of the Mandatory Rules makes special provisions for cases where a roof space in an existing two-storey dwellinghouse is to be converted to provide not more than two habitable rooms in the roof. The rules in this Appendix are mandatory for this case. However, when the conversion of the roof space of a two-storey dwellinghouse will provide more than two habitable rooms, or when the dwellinghouse already has three or more storeys before conversion of the roof space, Appendix B would not be relevant and the Code recommendations would apply.

Where a particular Code recommendation would be difficult or impossible to comply with for economic or practical reasons, consideration should be given to what compensating features, for example fire detection or automatic sprinkler systems, might be incorporated into the design to offset the deficiency. Such features could compensate for, for example, a lower than prescribed fire resistance in some part of the structure or a lack of full enclosure of an escape stairway. In such circumstances the local authority might be asked to consider granting a relaxation of, or dispense entirely with, the requirement which cannot easily be complied with. The preamble to the Mandatory Rules anticipates this course of action.

The Codes make few recommendations for structural fire protection since requirements B2 and B3 of the Regulations adequately control internal fire spread. In the case of Shops and Offices, the Fire Precautions Act 1971 may impose additional requirements at the fire certification stage.

In order to provide flexibility in design in respect of the use, size and arrangement of spaces within the building, the Codes explain the principles of fire safety and give general guidelines. Some specific limitations are placed on, for example, travel distances within different sections of the escape route. Diagrams in the form of floor plans, etc., are included in the Codes; they are intended to illustrate principles without being prescriptive.

The requirements for both structural fire safety and means of escape should be considered together in the early stages of the design process; means of escape provision is part of the more general problem of access and circulation. If this aspect of design is left to a late stage, radical changes may have to be made which may well affect other aspects of the design believed to have been finalised.

MANDATORY RULES: SECTION 1

Dwellinghouses

CONTENTS

MANDATORY RULES: SECTION 1

Dwellinghouses

INTRODUCTION

The Statutory Requirement
The requirement is that –

"... there shall be means of escape in case of fire from the building to a place of safety outside the building capable of being safely and effectively used at all material times."

Application of the Requirement
The requirement applies to the third storey and above in –

- a new building of three or more storeys which is or contains a dwellinghouse;
- an existing dwellinghouse which is extended or materially altered and will have three or more storeys when completed;
- an existing building of three or more storeys the use of which is changed to use as a dwellinghouse.

Paragraph 1.4 of the Mandatory Rules applies special requirements contained in an Appendix to –

- an existing two-storey dwellinghouse in which the roof space is converted to contain one or two habitable rooms.

Single and two-storey dwellinghouses are NOT subject to the requirement.

Note that, by definition, a dwellinghouse does not include a flat; note also that the committee which prepared Section 1.1 of BS 5588 specifically excluded from the scope of the Code "houses in multiple occupation".

The paragraph B1 in Schedule 1 to the Regulations does not require means of escape to be provided for the occupants of ground and first floors of dwellings; this is presumably because escape from these levels is hardly influenced by structural features of the building. If the stairway becomes unusable because of fire, it is usually possible for occupants to escape through windows. However, it must be noted that escape via windows is only possible if they are openable, large enough and suitably situated. In buildings of three or more storeys the protection of the escape route required for the third and higher storeys will also benefit the occupants of lower floors.

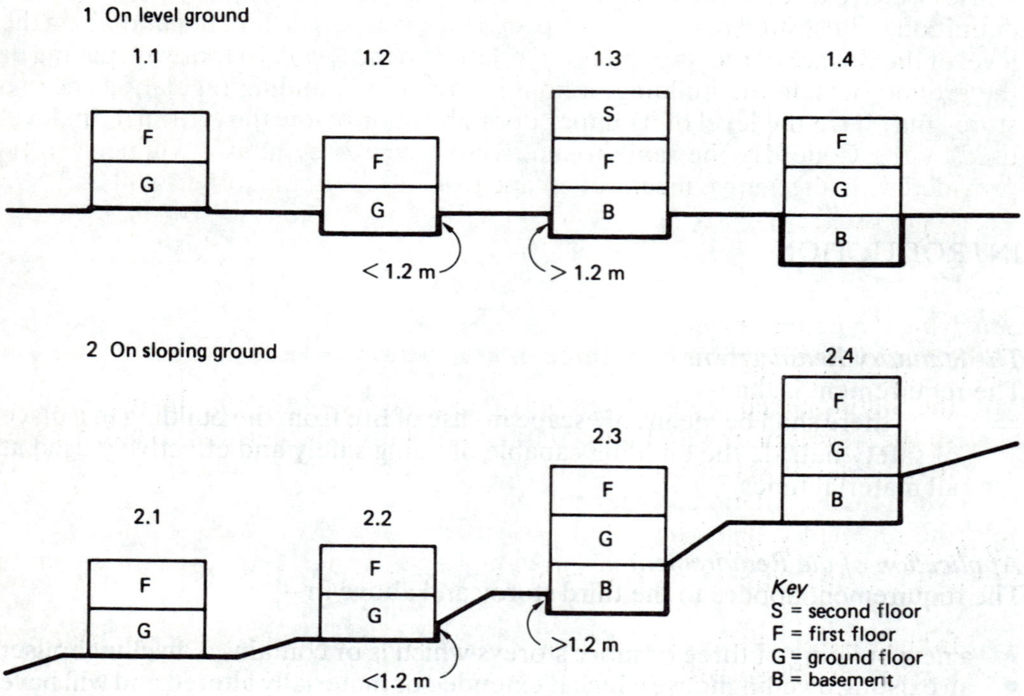

Notes

The Regulations require means of escape to be provided for dwellings with more than two storeys. Regulation 2(3) states that a basement is not to be regarded as a storey. The same Regulation defines a basement as a storey of which the floor is at any point more than 1.2 m below the finished surface of the ground adjacent to it. There is no definition in the Regulations of "ground storey"; the Code defines it in terms similar to the 1976 Regulations, "a storey the floor of which is situated at such a level or levels that any given point on its perimeter is at or about but not more than 1.2 m below the level of the finished surface of the ground adjoining the building in the vicinity of that point".

Figure 1.1(a) shows outline sectional elevations of buildings on level and sloping ground; the floors are identified in terms of the definitions. All are "two-storey" and the Mandatory Rules do not apply. Although Case 1.3 has a nominal second floor, it remains by definition a two-storey house.

Case 2.4 is an interesting one; the Mandatory Rules do not apply because it is technically two-storey and the upper floor is less than 3 m above ground level at the rear. However, in a fire, it might not be possible for those in rooms at the front to reach the rear of the building. At the front the floor is some 5 m above the level hard-standing. This might be considered justification for providing a protected internal escape route.

FIG.1.1.(a) Ground floors and basements

For the purpose of counting storeys, the definition of basement in Appendix A states that a basement is not to be regarded as a storey. Clause 5.5 of the Code makes recommendations for means of escape from any basement which contains a bedroom. The Mandatory Rules define a basement as a storey the floor of which is at some point more than 1.2m below the finished surface of the ground adjacent to it (see Fig. 1.1(a) (opposite).

The Code defines 'height of house' by reference to the highest point of the floor of the highest storey. Recommendations are then made for houses less than 4.5m high, houses which exceed this height by one floor level and finally, for houses which exceed 4.5m in height by more than one floor level. The effect of this definition is illustrated in Fig. 1.1(b) (page 18) which shows that, depending on the level of the surface of the 'ground floor' relative to the level of a footway, paving or the ground outside the building, a house defined by building regulations as two storey may have the level of its upper floor above or below the critical 4.5m level used by the Code. By the same token, some four storey houses will need to be provided with alternative means of escape from upper floors, others will not.

Satisfying the Requirement
In the case of dwellinghouses of three or more storeys –

(a) the Statutory requirement can only be satisifed by constructing the building so that it complies with the recommendations of certain clauses of BS 5588: Section 1.1: 1984, including any amendments incorporated up to July 1985 when the Regulations came into effect.
(b) the relevant clauses are listed in Section 1 of the Mandatory Rules. Notes explaining these clauses and providing additional information are given in the remainder of this Section.

In the case of a loft extension providing one or two rooms in the roofspace of a two-storey dwellinghouse –

(c) as an alternative to treating the extended dwellinghouse as a house of three storeys, work described in Appendix B of the Mandatory Rules may be executed. Explanatory notes and information relating to Appendix B will be found at the end of this Section.

DEFINITIONS AND TERMS

1 Concerning the Building

DWELLINGHOUSE

A building which contains a dwelling which is not a flat or maisonette. It should be noted that the Code uses the phrase "single-family dwelling" and states specifically that the recommendations do not apply to houses in multiple occupation. Only when the dwellinghouse has three or more storeys is the designer required by the Mandatory Rules to follow the recommendations of the Code.

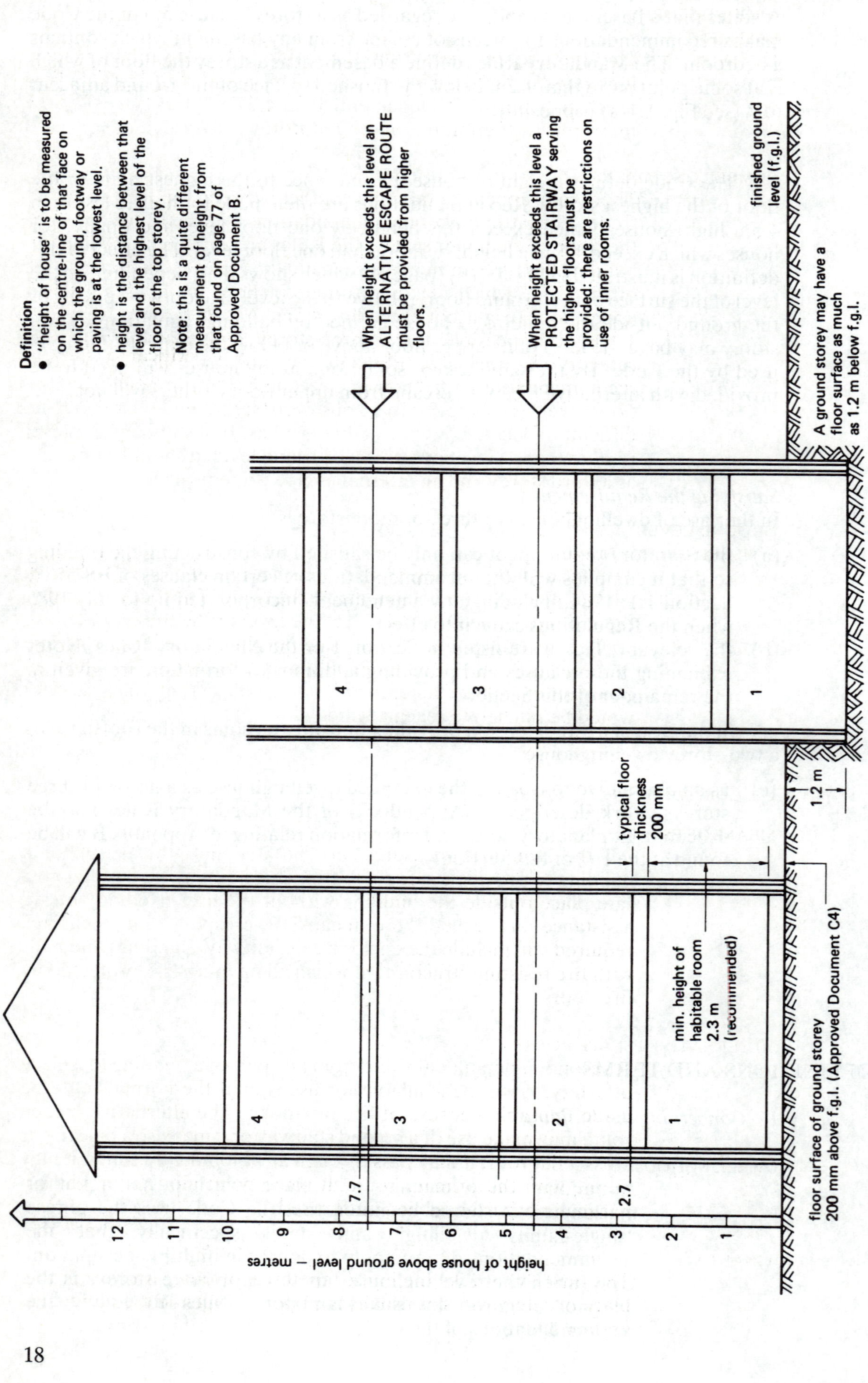

FIG.1.1(b) Definition of "height of house"

HEIGHT OF HOUSE
GROUND STOREY
BASEMENT

The method of measuring the height of a dwellinghouse is illustrated in Fig. 1.1(b) (opposite). This method differs from that used in Approved Document B2/3/4. Fig. 1.1(a) (page 16) interprets the definitions of ground storey and basement.

NUMBER OF STOREYS

There may appear to be some conflict between the Rules and the Code: The latter makes recommendations in terms of the height of the floor of the upper storey; the Rules are applied only when the building has three or more storeys and the Rules exclude clause 5.2 of the Code which makes recommendations for houses with an upper floor height not exceeding 4.5m. As can be seen from Fig. 1.1(b) it is possible for a three-storey house to fall into this category. In this case it would be wise to follow the Code recommendations for "houses exceeding 4.5m in height by one floor level" (clause 5.3). In a house with split floor levels it may be difficult to decide how many storeys it has. If any upper floor level is more than 4.5m above finished ground level, it should be treated as a third storey and an internal escape route provided.

GROUND LEVEL
FINISHED GROUND LEVEL

The terms "ground level" and "finished ground level" occur in various places in this document and the Code. The first is, literally, the ground surface; the second is the level of the top of any finish applied to the ground. What is important for means of escape purposes is the level, be it of soil, paving, or terrace, to which an occupant might have to jump or be lowered, or from which a ladder would be pitched to reach a window.

2 *Concerning the Escape Route*

MEANS OF ESCAPE

All the structural measures incorporated into the design of a building to provide a safe route from any point in the building to a safe place outside the building without the user needing outside assistance are called "the means of escape". The measures required will include the enclosure of stairways, landings and halls with fire resisting structure, in which all openings are protected by fire doors.

ALTERNATIVE ESCAPE ROUTE

A second, independent route from any part of a building to a place of safety; it will be available for use should the normal route be made unusable because of fire or smoke. The alternative escape route may make use of a second stairway or it may use a balcony or cross a flat roof; it may pass through an adjoining dwelling: it may merge with the primary route at some point along it at which a particular hazard has been outflanked.

FINAL EXIT

The final exit marks the end of the escape route; it gives on to a place of safety which is usually the open air outside the building at ground level.

PROTECTED STAIRWAY

A stairway forming part of an escape route, including any hall or passage between the foot of the stair and the final exit; it is separated from the remainder of the building by fire resisting construction and fire doors. The walls and floors enclosing the stairway must have a prescribed period of fire resistance (except when they are external walls) and doorways penetrating the enclosure will have self-closing fire doors fitted in them (see Figs. 1.2 and 1.5).

More information about protected stairways will be found in clauses 1.44 to 1.46 and Appendix E of Approved Document B2/3/4.

FIRE DOOR

A door which, when closed, will resist the passage of fire and hot gases. The door, the surrounding frame and the door furniture contribute to the fire resistance of the door. The Approved Document B2/3/4 contains information about fire doors in Appendix F and in item 9 of Table A1.

INNER ROOM
ACCESS ROOM

An inner room is a room which does not open directly off a hall or landing. An occupant of an inner room can only escape from it by passing through the access room. Examples of inner rooms and access rooms have been included in Figs. 1.2 and 1.5.

FIRE AND FIRE GROWTH

Fire Hazards in Dwellinghouses

The Code states:

"The only sound basis for designing means of escape from fire is to attempt to locate the position of all possible sources of any outbreak and to predict the course which might thereafter be followed by the fire as it develops or, more particularly, the routes which smoke and hot gases are likely to take."

To this advice it would be sensible to add:

"and to consider the likely behaviour of occupants of different ages and levels of responsibility when they find themselves threatened by fire".

Common sources of fire in the home are –

- An open fire: the absence of a fire guard to prevent falling embers; furniture placed too close to this source of radiant heat.
- The airing of clothes over a heat source not designed for this purpose.
- The storage of combustible liquids and gases in dangerous amounts or in unsuitable places.
- Radio and TV sets left connected to the mains.
- Cigarettes and matches carelessly discarded; children playing with matches; people smoking and falling asleep.
- The use of liquid fuelled portable heaters where they can be overturned by children or animals.
- Poorly maintained electrical apparatus, particularly those with cords and flexes; poorly maintained electric blankets.
- Cooking activities, especially when they involve frying in deep fat.

It can be seen from this list that –

- fire can start in any habitable room, or in a kitchen,
- fire can start in a room whether it is occupied or not,
- fire can start when occupants are awake and alert, or when they are asleep.

Thus, rooms and spaces in a dwelling may conveniently be classified as:

High hazard:	Living room, dining room, kitchen, bedroom, dressing room.
Low hazard:	Bathroom, WC, utility room, storage room.
Protected:	Entrance hall, inner hall, stairway, landing.

The recommendations of the Code only provide adequate safety on the assumption that potential sources of fire are excluded from protected circulation areas. The most serious hazard would arise if a fire were to start in the entrance hall of a dwelling from which the entrance door was the only exit. Unfortunately, it is in the entrance hall that portable heaters are placed when occupants find that the heating system provided is either inadequate or too expensive for them to run.

The Growth and Spread of Fire

The growth and spread of a fire once ignited is affected by –

- *The type and disposition of combustible materials* in the form of furniture and furnishing fabrics around the point of origin: some fabrics and furniture ignite more readily than others, have a high combustible content and some emit large volumes of toxic gases or smoke as they burn. Requirement BR. B2 controls wall and ceiling lining materials with the aim of slowing the growth and spread of fire within the room of origin.
- *The supply of oxygen to the fire*: Open doors or windows or the existence in the room of a flue all allow air (hence oxygen) to feed the fire and thus influence the rate of growth. In a well-sealed room a fire will use up the available oxygen and the burning rate will be retarded. However, if a door is then opened, unburned combustible gases are likely to "flash over" and the fire will rapidly revive to greater intensity; spread to other parts of the dwelling may then be rapid through the opened door.
- The fact that *smoke and hot gases are buoyant* and therefore spread most readily upwards: heat and smoke rise quickly to the ceiling of the room of origin where they spread sideways and fill the room from the top downwards. If the door is open they soon reach the hall or landing; then up the stairway to the upper floors. For this reason a room or dwelling entered down a flight of stairs from the floor above presents a greater hazard than one with its entrance door at the same level as the accommodation.
- It is important to *consider the kind of occcupant* for whom the dwelling is being provided. Certain human characteristics influence the reaction of a person to a fire incident. Age, physical fitness and mental responsibility are prime factors. Normally, people who are awake and in the room of origin of a fire will react quickly; if awake but in another room, awareness of the outbreak will be delayed and they may be faced with smoke in the circulation areas when they do attempt to escape. The very young and the old are more likely to behave irrationally and the old are less physically active. The behaviour of a disabled person may be constrained. Unfamiliarity with surroundings reduces the confidence of a person who has to escape through thick smoke and he or she may quickly become disorientated.

If fire breaks out when occupants are asleep there may be a delay before they become aware of danger. On waking they are likely to be confused at first and may panic. If the fire is in the bedroom, getting out will present less of a problem than if the fire starts in the living room or kitchen after the occupants have gone to bed. It is important that escape is made as easy as possible. For this reason it is recommended that stairways are enclosed and discharge at the foot reasonably near the front door and that all doors opening into the stairway enclosure are self-closing fire doors.

In the ground and first storeys of a house openable windows provide an alternative escape route should the hall and unenclosed stairway be unusable.

RECOMMENDATIONS FOR MEANS OF ESCAPE

Recommendations for Three-Storey Dwellinghouses

The Mandatory Rules require three-storey houses to be constructed in accordance with the recommendations of clause 5.3 of the Code; this deals with ". . . houses exceeding 4.5m in height by one floor level".

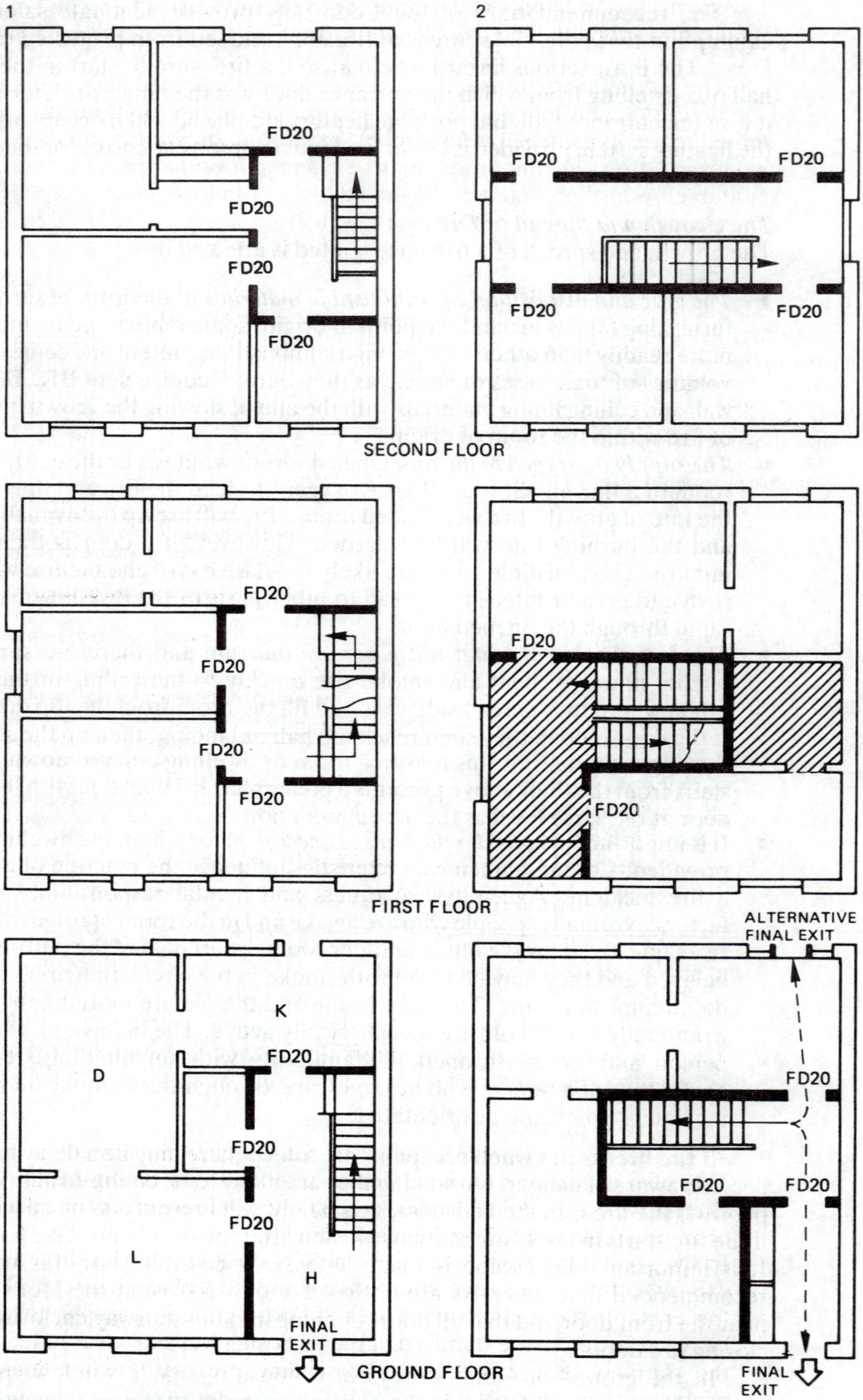

FIG.1.2 Two design options in three-storey dwellinghouse

There are two design options. In the first, the upper storey is separated from those below by fire-resisting structure (a compartment floor, fire-resisting walls where necessary and fire doors across points of access). The storey is then provided with its own, independent means of escape. The second option is to secure the escape route from the upper floor by using the internal stairway. The stairway must be constructed as a "protected stair" with fire resisting enclosure down to the ground storey and continuing to the final exit. The second option is the one usually adopted. The requirements for protected stairways are shown in Fig. 1.2.

The recommendations of the Code are also concerned about "inner rooms" (see definition and Fig. 1.2). If they occur at all on the top floor they should be limited to use as a kitchen, utility room, bathroom, dressing room or WC. On the two lower storeys an inner room may be planned for other uses (for example, as a habitable room) provided it is served by a window which may be used in emergency for escape or rescue (see Fig. 1.3).

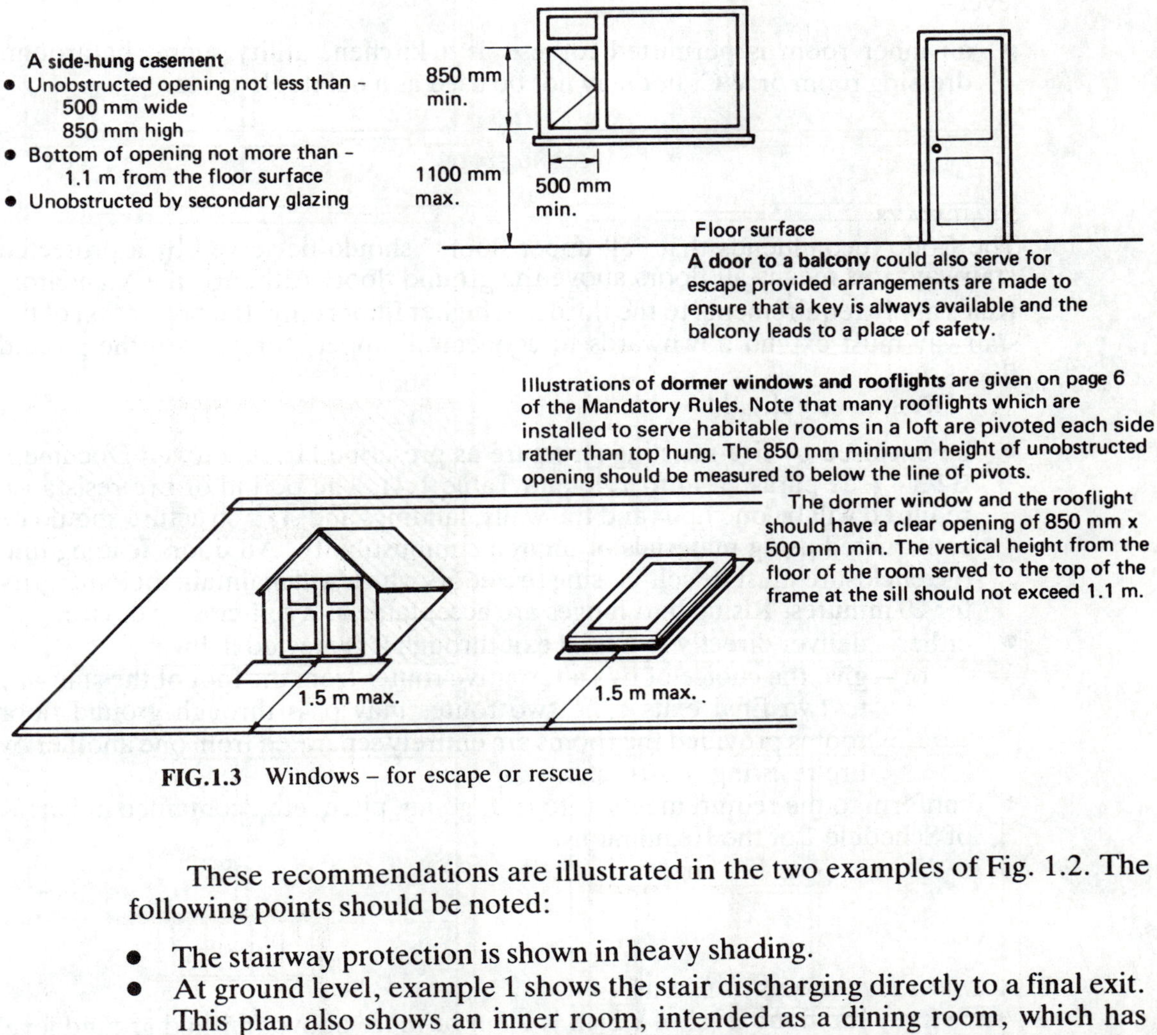

FIG.1.3 Windows – for escape or rescue

These recommendations are illustrated in the two examples of Fig. 1.2. The following points should be noted:

- The stairway protection is shown in heavy shading.
- At ground level, example 1 shows the stair discharging directly to a final exit. This plan also shows an inner room, intended as a dining room, which has alternative escape routes through the kitchen or lounge and the added security of an openable window or french door.
- In the second example, a central stairway does not deliver directly to a final exit but offers instead two alternative routes to a place of safety. Note also in this example the cross-hatched area of floor on the first storey; this area must be fire resisting to the standard of the enclosing walls to ensure the integrity of the stairway. For this reason the scheme presents greater constructional problems than example 1.

Recommendations for Dwellinghouses of more than Three Storeys

Dwellinghouses of four or more storeys must comply with the recommendations of the Code in relation to –

(a) the use and planning of inner rooms;
(b) the protection of stairways;
(c) the provision of alternative means of escape.

INNER ROOMS

In any storey which has a floor level not more than 4.5m above finished ground level –

- an inner room is permitted for any use provided it has a window which can be used for escape in the event of fire (see Fig. 1.3).

In any storey which has a floor level more than 4.5m above finished ground level –

- an inner room is permitted for use as a kitchen, utility room, bathroom, dressing room or WC, but may not be used as a habitable room.

STAIRWAYS

The Code recommends that "all upper floors" should be served by a protected stairway; this means all floors above the ground floor. Although the Mandatory Rules apply requirements to the third and higher floors only, the protection of the stairway must extend downwards to connect all upper storeys with the ground floor.

The stairway should –

- be enclosed by fire-resisting structure as prescribed in Approved Document B2/3/4 (see paras. 1.44 to 1.46 and Table 1.1). The period of fire resistance required will be one hour and the walls, landings and stair structure should be constructed using materials of limited combustibility. All doors leading into the enclosure must be self-closing fire doors which will maintain their integrity for 20 minutes. Rising butt hinges are acceptable as a self-closing device.
- either – deliver directly to a final exit through a protected hallway;
  or – give the choice of two alternative routes from the foot of the stairway to two final exits. The two routes may pass through ground floor rooms provided the rooms are entirely separated from one another by fire resisting construction;
- conform to the requirements as to rise, going, pitch, etc., contained in Part K of Schedule 1 of the Regulations.

ALTERNATIVE MEANS OF ESCAPE

Any storey the floor level of which is 7.5m or more above finished ground level (see Fig. 1.1(b)) must be provided with an alternative escape route. This may make use of a second protected stairway within the building, or a balcony, or a flat roof, provided it is independent of the primary route and leads to a place of safety at ground level.

The stairway which in the downwards direction provides the primary escape route may also be used upwards as the alternative means of escape if, at the top, it connects with another route to the ground and provided it is constructed so as to prevent fire spread between storeys (see Fig. 1.4).

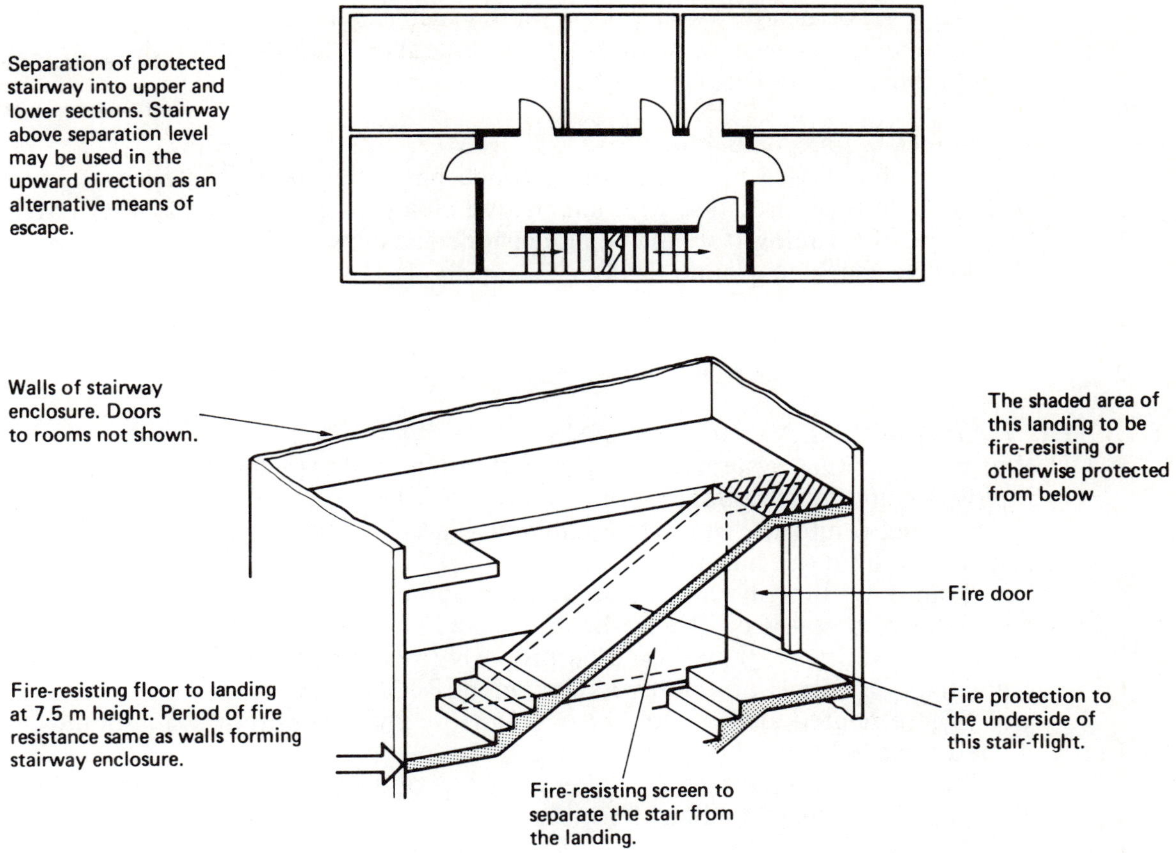

FIG.1.4 Protection of stairways

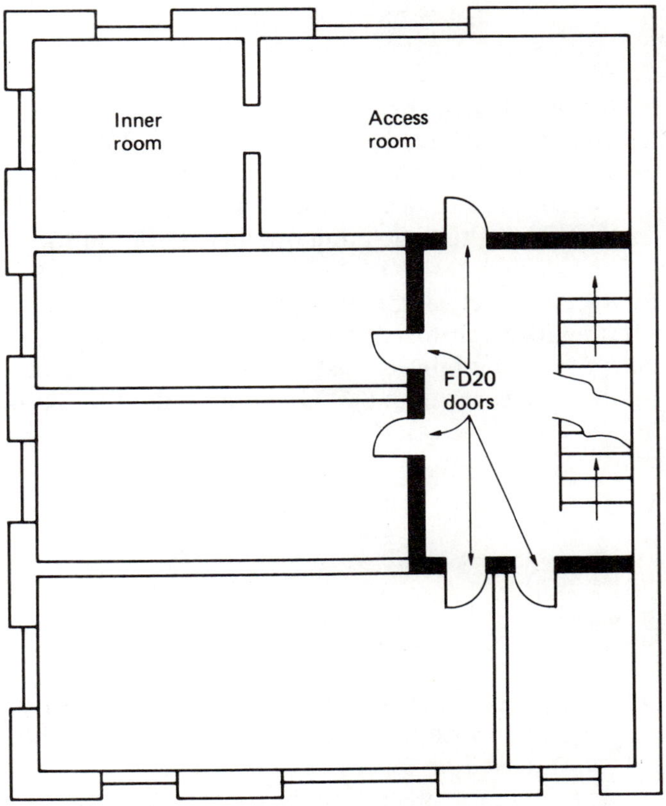

FIG.1.5 Inner rooms and protected stairways

Inner room

Definition: a room which does not open directly off a hall or landing. Escape must be made through an access room (or rescue by way of a window).

- an inner room may be used as a kitchen, utility room, bathroom, dressing room or WC on any storey irrespective of height of the uppermost floor;
- an inner room should not be planned as a habitable room on any storey the floor of which is more than 4.5 m above ground (see Fig. 1.1(b));
- an inner room may be planned as a habitable room on a storey the floor of which is less than 4.5 m above ground provided it has a window suitable for escape or rescue (see Fig. 1.3).

Protected stairways

The landing of all upper floors must be within the enclosure. At the foot the hall or passage leading to the final exit to be similarly enclosed.

The period of fire resistance for enclosing structure is –
in a three-storey dwellinghouse – half an hour;
if four or more storeys – one hour.

If the period is one hour, construction materials are required to be "materials of limited combustibility" (Approved Document B2/3/4, Appendix A).

All openings in enclosing walls protected by FD 20 fire doors.

The foot of the stairway delivers either –
directly to the final exit; or
offers two or more separate routes to different final exits,
one of which must be protected.

Any part of an alternative escape route which uses a balcony or flat roof must be guarded in accordance with the requirements of K2 of Schedule 1 of the Regulations; BS 6180 also gives guidance.

Recommendations for Basements

If any dwellinghouse has a basement which has a bedroom located in it, that bedroom must be provided with alternative means of escape. If the alternative makes use of a window it should be dimensioned as shown in Fig. 1.3 and the route should give access to ground level.

DUCTED WARM AIR HEATING SYSTEMS

Warm air heating systems (or air conditioning systems) can circulate smoke rapidly around the building; this would be particularly serious were the system to spread smoke into the protected stairway. The Code makes recommendations intended to limit this hazard.

Transfer grilles should not be fitted in any element forming the protective enclosure of a stairway. Where heating ducts pass through the walls or floors enclosing a stairway, they should be properly firestopped. Reference should be made to Appendix H of Approved Document B2/3/4 in this respect.

Where heated air is ducted into a stairway enclosure, return air should be ducted directly back to the heater.

Warm air terminal points – grilles, registers or diffusers, and return air grilles should be positioned at a height not exceeding 450mm above floor level.

A room thermostat should be located in the living room, positioned at a height above floor level of between 1370mm and 1830mm and set to prevent the maximum temperature exceeding 27°C.

FIRE DOORS

In a dwellinghouse, fire doors associated with the means of escape will only be required to protect openings in the structure enclosing the stairway. For these doors the Code recommends a minimum performance of FD 20. This means a door which maintains integrity for 20 minutes when tested in accordance with BS 476: Part 8. (The designation FD 20 has this meaning on illustrations in these notes.)

Fire doors should be fitted with a self-closing device, capable of closing the door over any angle of swing and against the resistance of the latch; rising butt hinges may be suitable for this purpose. Door hinges must be non-combustible and the metal must have a melting point not lower than 800°C; aluminium hinges are not adequate.

GLAZING

If areas of glazing are used in walls or doors which must be fire resisting the following restrictions apply:

(a) if the glazing is fire resisting for the appropriate period in terms of integrity, stability and insulation, it may be used without restriction. Glass is available which achieves this standard for at least 30 minutes;
(b) glazing which can satisfy the test only in respect of integrity and stability may be used for fire resisting screens within stairway enclosures.

LOFT CONVERSIONS IN TWO-STOREY DWELLINGHOUSES

Introduction

Requirement B1 of the Building Regulations requires effective means of escape, which can be safely and effectively used at all material times, to be provided. One class of building to which this requirement applies is a two-storey dwellinghouse extended or altered to add a third storey. The means of escape need only afford escape for people occupying the third storey.

Appendix B of the Mandatory Rules makes requirements specifically for loft conversions in which the third storey is intended to provide not more than two habitable rooms. For anything larger than this the recommendations of the Code of Practice (BS 5588: Section 1.1: 1984) would have to be followed. The Appendix provides a definition of habitable room in this context. As well as the usual living room and bedroom, the definition includes a kitchen, but excludes a bathroom. Appendix B could therefore be used where the loft conversion comprised, say, two bedrooms and a bathroom.

The requirements concern the stairway serving the new storey, the enclosure of that stairway and changes to be made to doors on the ground and first storeys which separate habitable rooms from the stairway on those two levels. The requirements also call for facilities for emergency window rescue from the loft area in case the stairway should become unusable.

It should be noted that under the Town and Country Planning General Development Order 1988, planning permission is now required for any loft extension which includes an alteration to any part of the roof. It can be assumed that this only applies if the alteration changes the external appearance of the roof.

Fire Resistance

The definition of "Fire Resistance" prescribes the periods to be achieved by doors and enclosing structure wherever these elements are required to be fire-resisting.

DOORS

A door must have 20 minutes integrity: In the Code this would be designated an FD 20 door.

ENCLOSURES

Walls enclosing stairways (and a floor if it forms part of the enclosure) must achieve the full 30 minutes stability, integrity and insulation.

Approved Document B, Appendix A, para A2 and Table A1 give more information about fire resistance of elements and the meaning of terms.

The Requirements of Appendix B

1 STAIRWAYS

Depending on the size and layout of the existing first storey of the two-storey dwelling, the location of the new stairway to the loft will vary. The requirements in Appendix B give three possibilities:

1.1 The new stairway may be accommodated above and within the existing stairway enclosure. In this case a self-closing fire-resisting door must be located at the top of the new stairway, separating the stairway from the new rooms in the loft. This door must be set in a fire-resisting enclosure; this implies that existing stairway-enclosing walls will be carried upwards to meet the roof. This ensures that should smoke from a fire occurring on a lower storey render the stairway unusable, those in the loft are protected until they can be rescued via windows.

1.2 The new stairway may start from the existing stairway enclosure at first-storey level but be carried upwards through an existing room. In this case the new stairway must be separated from the room and the remainder of the dwellinghouse by fire-resisting enclosure. A self-closing fire-resisting door must be located either, as in example 1.1 at the top of the flight, or at the bottom of the new stairway.

1.3 The new stairway may have to be located wholly in an existing first storey room. In this case the entire room must be separated from the remainder of the dwelling by fire-resisting enclosure and there must be a self-closing fire-resisting door at the top of the new stairway as in example 1.1.

Problems with stairways. It is frequently difficult, if not impossible, to fit a normally pitched stairway into the available space. For this reason approval has been sought for stairways pitched steeper than 55°: Approved Document K defines these as ladders. The Secretary of State has been asked for several Determinations on this matter (Determinations 7, 26, 29 and 41 BR*) and has recently issued a guidance note (reproduced at end of these notes) giving his Department's view. His advice is that steep stairways are acceptable provided they can be used safely and give access to only one habitable room: they should have handrails on both sides to compensate for the steepness. A proprietary make of steep stairway has been accepted by some authorities.

Appendix B appears to assume that existing stairways will be fully enclosed within the ground and first storeys. This is not always the case. A number of modern designs have the stairway rising from an open plan living area on the ground storey and the landing in the first storey may take the form of a balcony overlooking the living room. This type of layout provides a very unsatisfactory means of escape for the occupants of a converted loft. At present, nothing in the mandatory rules prevents such a layout being accepted.

2 UPGRADING DOORS ON THE GROUND AND FIRST STOREYS

All doors separating the stairway enclosure from habitable rooms are required by the mandatory rules to be modified as follows:

On the FIRST STOREY existing doors should be fitted with self-closing devices; rising butt hinges are suitable. At the same time any glazing in these doors should be replaced with wired glass. (Although Appendix B does not require it, it would be prudent to treat any other glazing, e.g. between the hall or landing and a habitable room providing borrowed light, in a similar manner.)

On the GROUND STOREY–
- if only one habitable room is provided in the loft – existing doors should be treated as on the first storey;
- if two habitable rooms are provided in the loft – existing doors should be replaced with self-closing fire-resisting doors (FD 20).

Where an opening through stairway-enclosing walls is not fitted with a door, nothing in the Mandatory Rules requires a door to be fitted.

3 WINDOWS AS AN ALTERNATIVE MEANS OF ESCAPE

Appendix B requires that "the room shall have a window or rooflight . . .". Thus, if two habitable rooms are provided on the new upper storey, it would appear from this form of words that each must have a window or rooflight. It might be argued that, with a fire door at the head of the stairway protecting the entire loft area, one window or rooflight would be sufficient.

The window or rooflight opening must have minimum dimensions of 850mm high and 500mm wide. These dimensions refer to the clear open area through which a person may climb in an emergency. The window or rooflight must be so positioned that the bottom of the opening is not more than 1.1m above the floor level of the room (no minimum height is given) and, externally, not more than 1.5m from the eaves (where a ladder can be pitched) (see Fig. 1.3).

These two limiting dimensions have resulted in some windows being set with the cill close to floor level, particularly when the roof has a low pitch. A window close to the floor may be awkward to use in an emergency and creates a different

---

*These refer to numbered selected Determinations published, with annotations, in "Knight's Building Regulations (with Approved Documents)" – full details of which title can be found on the inside back cover of this book.

hazard, especially for children who may fall out. Requirement K2, coupled with the guidance in the corresponding AD, suggests that where any cill is lower than 900mm the window should be guarded, but normal guarding could make the window difficult to use. A simple calculation shows that with a minimum cill height of 900mm and a maximum of 1.5m down the roof from cill, the roof pitch cannot be less than about 37°.

This problem has been put to the Secretary of State for his determination (Determinations 23 and 27 BR). He also dealt with the matter in his guidance note (following). He has accepted a small increase (but does not quantify) in the 1.5m limit and favours the use of safety glass (laminated or toughened) for a window which is set with the cill closer than 900mm to the floor. Despite these rulings, problems remain and an alternative solution is needed.

Another Determination of interest (BR number 8) concerns a window located in the gable wall, rather than in the roof slope as is shown in Fig. 1 of Appendix B. The Secretary of State determined that a window in the gable wall was satisfactory as a means of escape provided it could be reached by ladder from outside. He stated that Appendix B did not exclude a window in a gable.

Department of the Environment Guidance on Loft Conversions

"STAIRS OF UNUSUAL DESIGN AND STEEPNESS

1. A problem has arisen as a result of a recent trend in loft conversion design. Stairs which have shaped treads of a design which has not previously been used in this country to any great extent are now being installed more frequently. The pitch at which these stairs are installed brings them within the definition of "ladders" contained in the Approved Document supporting Part K of the Regulations on stairs (i.e. they are steeper than 55°). For this reason they are not covered by the guidance given in the Approved Document on stair dimensions, although it is considered that other provisions that relate to stairways (e.g. headroom, landings and guarding) can still be applied.

2. In the absence of specific evidence about the safety of these stairs, it is the Department's view that they are capable of providing adequate means of escape as required by Part B1 of the Regulations in certain limited use conditions. These are as access to a single room loft conversion where space is not available to provide a stair designed in accordance with the guidance given in relation to Part K. It is however considered that, because of their steep pitch, handrails on both sides of the stairs are essential.

3. Evidence about the safety of installations of this nature is now being collected, and will be taken into account when the means of escape provisions are being revised. Consideration will also be given to the desirability of bringing such stairs within the scope of the guidance supporting Part K.

OPENABLE ROOF WINDOWS AT LOW LEVEL

4. A second problem relates to the trend towards the use of openable roof windows as a means of providing emergency egress and rescue from loft conversions. The criteria in the means of escape rules include a limiting distance of 1.5m from the bottom of a window opening to the point at which a ladder for rescue purposes might be pitched. Whilst a maximum height to the cill of windows from inside the room is specified in the rules, there is no minimum distance given. In a number of cases, this has resulted in roof windows being set at a very low level, and sometimes almost at floor level. As the window needs to be readily openable for escape purposes, this leads to the risk that it may become a hazard in day-to-day use, especially in cases where small children are in the room. Apart from the general safety issue, it is also questionable whether a window set almost at floor level is suitable for means of escape purposes.

5. In a number of recent cases, it has been decided that low level cills do not adequately meet the requirements of Part K2 of the Regulations relating to protection from falling unless the edge of the floor is guarded. In addition, we are considering the need for an amendment to the means of escape requirements so that they include a minimum as well as a maximum dimension from floor to cill height for an emergency egress window. Alternatives to a roof window set at very low level should therefore be found where possible. A slight increase in the 1.5m dimension may be acceptable in some cases, particularly if this enables any risk to safety to be minimised. Another possible solution in some cases is a dormer window. It may also be possible to meet the means of escape requirements in some other way, so that there is no longer any need to rely on the use of a window.

PROTECTION OF STAIRWAYS FOR MEANS OF ESCAPE

6. Finally, the means of escape rules do not at present specifically require the stairway to be protected in situations where there may not be a stairway enclosure in the existing house. The Department's Stage 2 consultation paper issued in December 1986 proposed that such a requirement should be included, and we intend to implement this proposal when Part B of the Regulations is amended. In the meantime, the view that has been taken is that, where the criteria in the mandatory rules are fully met, a proposal must be accepted, even though the stairway may be open at ground-storey level, or there may be openings in the existing stairway enclosure that do not have doors in them.

Department of the Environment
June 1988"

APPENDIX
Relevant clauses of BS 5588: Section 1.1: 1984

The following clauses of the BS Code of Practice BS 5588: Section 1.1: 1984 are cited in para. 1.3 of Section 1 of the Mandatory Rules. They are listed here together with an indication of the subject dealt with by each clause.

SECTION 2 Definitions

SECTION 4 Planning in relation to fire
- 4.1 General
- 4.2 Fire development within a dwelling
- 4.3 Fire development originating outside the dwelling
- 4.4 Fire brigade access

SECTION 5 Means of escape and provisions for rescue
- 5.1 Commentary
- 5.3 Recommendations for houses exceeding 4.5m in height by one floor level
- 5.4 Recommendations for houses exceeding 4.5m in height by more than one floor level
- 5.5 Recommendations for basements
- 5.6 Recommendations for escape by way of an external balcony or flat roof
- 5.7 Recommendations for windows for escape or rescue purposes

SECTION 6 Construction
- 6.1 General
- 6.2 Protection of structural members
- 6.3 Fire resistance
- 6.4 Glazing
- 6.5 Fire doors
- 6.6 Recommendation for stairways (including helical stairways)

SECTION 7 Heating
- 7.3 Ducted warm air heating systems

MANDATORY RULES: SECTION 2

Flats and Maisonettes

in a building of three or more storeys

CONTENTS

MANDATORY RULES: SECTION 2

Flats and Maisonettes

in a building of three or more storeys

INTRODUCTION

The Statutory Requirement
The requirement is that –

> ". . . there shall be means of escape in case of fire from the building to a place of safety outside the building capable of being safely and effectively used at all material times."

Application of the Requirement
Schedule 1 of the Regulations applies this requirement to a new building which is of three or more storeys and which contains a flat. The Mandatory Rules, Appendix A, define a flat as "separate and self-contained premises constructed or adapted for residential use and forming part of a building from some other part of which it is divided horizontally". The phrase "divided horizontally" implies the existence of floors separating dwellings (that is, "party floors").

The term "maisonette" is not used in the Regulations but is defined in the Code. It can be assumed that, in the Regulations, maisonettes divided horizontally from one another are the same as flats. In the Code flats are dealt with separately from maisonettes because in a flat the living room, bedrooms and the kitchen are all on the one floor; in a maisonette, bedrooms are usually not on the same floor as the living room and kitchen. These layout differences are significant when planning satisfactory means of escape.

In the introductory remarks to these notes on means of escape (page 10), it was held that the Mandatory Rules do not apply to an existing building which is to be extended or materially altered to provide a flat or flats on the third or higher storey. Reasons were given for this interpretation. However, reference was also made to a DoE memorandum which puts forward a contrary view. If the DoE view is accepted, the recommendations of the relevant clauses of the Code will have to be followed.

Note that the Code was written for use in planning new blocks of flats (see the opening sentence of the Foreword to the Code). It will not always be possible to incorporate everything the Code recommends in an existing building undergoing material alteration or extension or subject to a change of use. In these circumstances it may be advisable to ask the local authority to relax or dispense with one or more requirements. To justify a relaxation, the local authority may suggest incorporating alternative active fire precaution measures.

Requirement B1 of Schedule 1 and thus the Mandatory Rules apply to a building which –

- is being newly erected,
- will have three or more storeys when finished, and
- will have a flat or maisonette on the third storey or above.

The requirement in B1 of Schedule 1 of the Regulations is also limited in its application to "escape for people from the third storey and above". However, the protected route provided for these people must extend through the lower storeys to a place of safety, outside the building, usually at ground level. Satisfying the Code recommendations places some constraints on the planning of the lower two storeys and on the position of features such as the boiler house, a garage or car park, access to a basement and any office for a porter.

Satisfying the Requirement
The requirement can only be satisfied by constructing the building in accordance

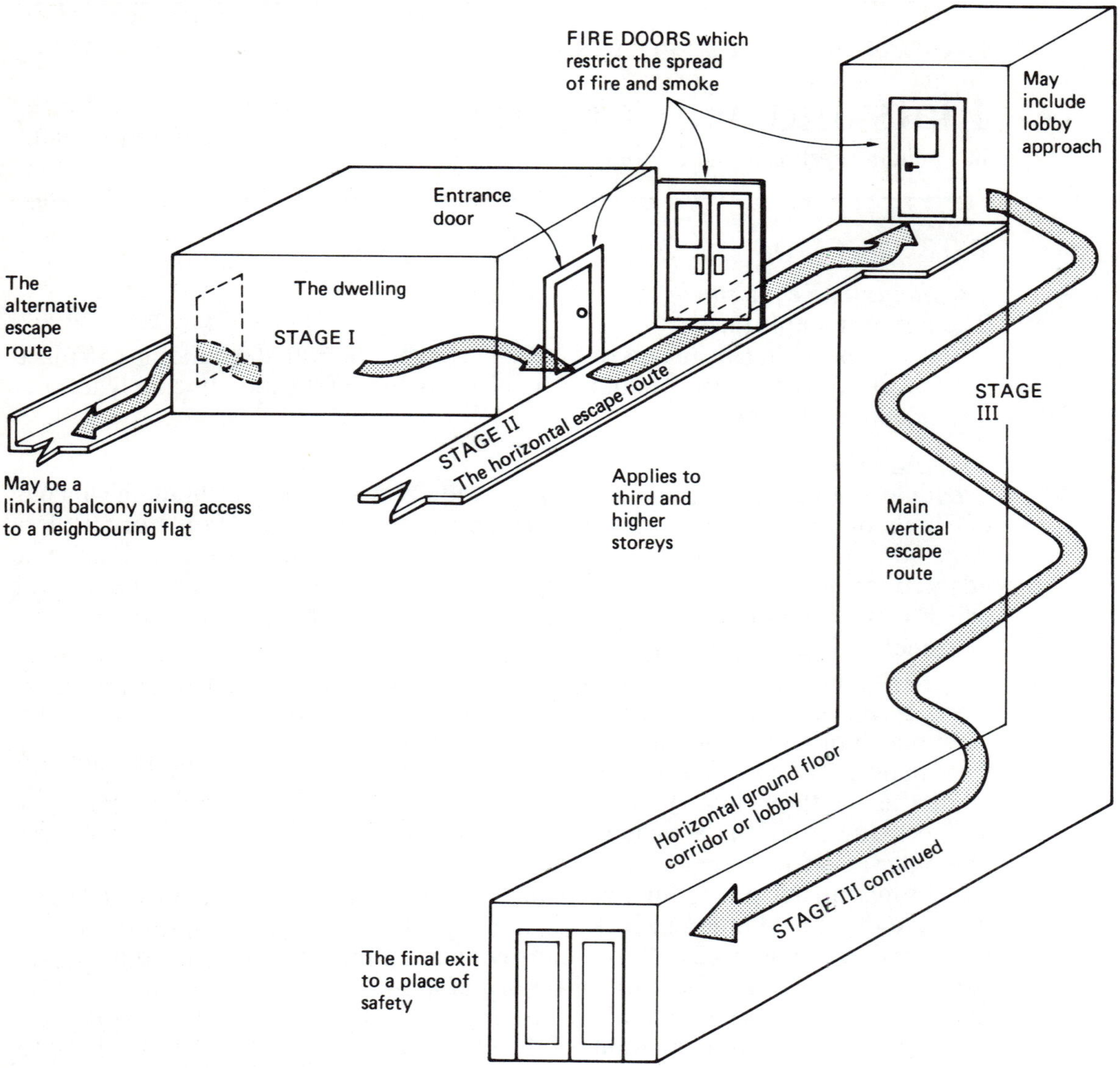

FIG.2.1 Stages of the escape route

with certain clauses of BS CP3: Chapter IV: Part 1: 1971, the Code of Practice for means of escape from flats and maisonettes in the event of fire. The version to be followed is that existing when the Mandatory Rules came into effect in 1985, including any amendments which had been made to the Code at that date. The relevant clauses are identified in the Mandatory Rules, Section 2. The relevant clauses of the Code are examined in these Notes; where appropriate, additional comment has been added.

General Considerations
The Code states: "The only sound basis for designing means of escape from fire is to attempt to locate the position of all possible sources of any outbreak of fire and to predict the courses which might thereafter be followed by the fire as it develops or, more particularly, the routes which smoke and hot gases are likely to take."

In flats and maisonettes the sources and causes of fire are the same as in dwellinghouses; the reader should refer to the paragraphs of Section 1 of these Notes headed "Fire Hazards in Dwellinghouses" and "The Growth and Spread of Fire" in which these matters are discussed.

A block of flats, constructed in accordance with the requirements of B3 of Schedule 1 of the Regulations, will have compartment walls and floors between dwellings. Flats will be similarly separated from common circulation areas. Stairways will be enclosed and other structural measures taken to control and restrict fire spread, including the requirements of B2. By these measures, *fire spread* beyond the flat of origin is unlikely. The main hazard arises from *smoke* entering corridors and stairways.

It is not usually necessary, if fire breaks out in a dwelling, for the entire building to be evacuated; the occupants of other flats should be safe if they remain where they are. Despite this assurance people will often prefer to leave the building and the means of escape provided should allow them to do so unaided and safely.

In a multi-occupancy building, where, for example, flats are over shops or associated with offices, the means of escape provided for the flats should be independent of, and quite separate from, the other occupancy. In no way should it be possible for the escape route from the flats to be threatened by fire in the other occupancy.

Stages of the Escape Route

This Code advises the designer to plan the means of escape for the occupants of dwellings in a block of flats in three stages as follows:

STAGE I – This stage concerns the escape of occupants from the flat of origin; the route extends from any part of the dwelling to the main exit from the dwelling. An alternative means of escape may be necessary in Stage I to achieve safety for all occupants of the dwelling; this may involve passing through an adjoining dwelling to reach the corridor and thus join Stage II.

STAGE II – This stage extends from the door of a flat along the horizontal route to a door leading on to a main stairway. The design objective in Stage II is to ensure the safety of occupants of flats on the same floor as the flat of origin.

STAGE III – This stage consists of the vertical route down a stairway linking Stage II to a final exit from the building. It provides for the safety of the occupants of dwellings on other floors should fire or smoke spread beyond the floor of origin.

It should be noted that these Stages interact and that solutions adopted for one stage may have implications for another. They are illustrated in Fig. 2.1.

The Planning of Dwellings and Blocks

The following features will influence requirements for means of escape:

- Approach to the dwelling
  - whether by internal corridor,
  - or external balcony or deck,
  - or by subsidiary stairways.
- Entrance to the dwelling
  - whether the same level as living rooms, or at bedroom level,
  - or via internal stairs from above or below.
- Internal arrangement
  - whether bedroom doors are nearer to the entrance door to the dwelling than are the doors of the living room or the kitchen.
  - whether in a maisonette the floor on which the living room and kitchen are situated is above or below the floor containing the bedrooms.
- Size of dwelling
  - large floor areas, which produce excessive travel distances between rooms and the entrance door, necessitate the use of inner halls and alternative escape routes, for example, by linking balconies.

- Provision of an alternative escape route
  - by way of a balcony leading directly to a stairway
  - or by way of an internal stairway leading up or down to an exit on another level,
  - or by way of a linking balcony through an adjoining dwelling to an exit.

Means of Escape Requirements
Safe means of escape is achieved by use of the following measures in the planning of dwellings and the approaches to them:

- The location of habitable rooms and kitchens relative to one another and to protected areas and exits.
- The provision of an alternative escape route from particular rooms or places.
- The enclosure of specified areas in fire resisting construction and the use of fire screens to protect escape routes.
- The provision of fire doors to contain the spread of smoke and fire; the use of stairway protection lobbies between corridors and stairways.
- Limiting the distance an occupant would need to travel through an area which could be affected by smoke.
- The use of either natural or mechanical ventilation to disperse smoke.
- By specifying the dimensions and construction of stairways to make them safe for use in an emergency.
- By ensuring that features which could impair the safety of escape routes are excluded from them; for example, storage rooms, boiler rooms and some service ducts.

DEFINITIONS AND TERMS

1 Concerning the Dwelling

DWELLING

A self-contained living unit which may be a flat or a maisonette.

FLAT

A dwelling which has all habitable rooms and kitchen on one level or, in the case of split level flats, not more than half a storey apart.

MAISONETTE

A dwelling which has habitable rooms and kitchen divided between two or more levels, the levels being more than half a storey apart.

DWELLING OF ORIGIN

The dwelling in which a fire starts.

ADJOINING DWELLING

A dwelling which adjoins the dwelling of origin; it may be on either side, or above or below. The term could be taken to include adjacent dwellings with entrance doors immediately across the access corridor.

ENTRANCE HALL

A protected space immediately inside the entrance door of the dwelling into which other room doors open.

INNER HALL

A protected hall or lobby into which room doors open but which is not directly connected with the entrance door.

LANDING

A circulation space at the head of a flight of stairs on to which rooms may open.

PRIVATE STAIRWAY

A stairway within a dwelling connecting different levels for use only by the occupants of that dwelling.

2 Concerning Common Areas and Access

BALCONY ACCESS

An external, open-sided balcony in common use giving access to a number of dwellings on one level.

DECK ACCESS

A wider version of balcony access.

CORRIDOR ACCESS

An internal, enclosed, common circulation space giving access to a number of dwellings.

FINAL EXIT

The end point of an escape route; the final exit discharges to a place of safety usually in the open air.

PROTECTED STAIRWAY

A common stairway, usually the main stairway, forming the vertical escape route to the ground storey. It is protected by fire resisting structure and approached at each level through self-closing fire doors.

SUBSIDIARY STAIRWAY

A stairway connecting not more than three flats with an access balcony (or deck) the flats being either on the storey above or the storey below the balcony.

MAIN STAIRWAY

A common stairway serving all storeys which forms the vertical escape route. It is required to be protected: see "protected stairway".

3 Concerning Means of Escape

ESCAPE ROUTE

The route from any point in a dwelling or other part of a building to a place of safety, usually in the open air at ground level. For convenience the route may be considered in three stages (see Introduction (p. 37) and Fig. 2.1).

ALTERNATIVE EXIT

An exit from a room, storey or dwelling which is additional to and separated from the normal exit.

ALTERNATIVE ESCAPE ROUTE

A route which the occupants of a room, storey or dwelling may use which is an alternative to all or part of the normal escape route. It is for emergency rather than everyday use.

DEAD END

Any area from which escape is possible in one direction only.

LINKING BALCONY

A small, private balcony linking alternative exits from two adjoining dwellings. It forms part of an alternative escape route.

STAIRWAY PROTECTION LOBBY

A space between a corridor and a protected stairway, separated from the remainder of the building by fire-resisting construction and guarded by fire doors; it gives the stairway increased protection from the entry of smoke and hot gases from the corridor. It is ventilated either permanently or by openable windows or ventilators.

FIRE-RESISTING DOORS

Doors which, complete with frame, hinges and furniture have been shown by prototype testing to BS 476 to be capable of checking the spread of fire and/or smoke for a specified period. They are fitted with a self-closing device and performance differences are indicated by Type Number (1, 2, 3 or 4).

VENTILATED LOBBY

A compact length of corridor or small circulation area which is permanently ventilated to the open air.

THE ESCAPE ROUTE: STAGE I – ESCAPE WITHIN AND FROM THE DWELLING

The Internal Planning of Flats

The main recommendations made by the Code with respect to the planning and construction of flats are set out below under key headings. The recommendations are illustrated in Figs. 2.2 and 2.3.

ENTRANCE DOOR

Where the flat is entered from an enclosed corridor, or ventilated lobby the entrance door should be fire resisting and self closing to inhibit the spread of fire from the dwelling into Stage II. Rising-butt hinges are sufficient. With all entrance doors, even those opening onto a balcony, the type of lock fitted needs to be chosen carefully; it should be openable from outside unless deliberately locked and it should be easy to open from inside without having to hunt for a key.

ENTRANCE HALL

All walls enclosing the entrance hall should have half-hour fire resistance; the living room and bedrooms should open directly on to the hall through doors which are fire resisting and self closing. Similar doors should be fitted between the kitchen and the hall. Any store rooms opening off the hall should also have fire resisting doors but they need not be self closing. All possible steps should be taken to prevent fire starting in the entrance hall or spreading rapidly into it. A bathroom or WC is not considered to present a hazard.

ENTRANCE ON ANOTHER FLOOR

Access to flats may be by way of a private or subsidiary stairway from the floor above or the floor below. A private stairway and any hall or landing inside the entrance door should be included within fire-resisting enclosure. Entry from the floor above presents the greater hazard; because smoke and hot gases rise, the stairway and upper landing are more easily made impassable. This could be fatal if fire developed in the living room while occupants were asleep. It is therefore a

Code recommendation that whenever entry is down a stairway from above an alternative escape route should be provided which can be used by the occupants of all bedrooms. Recommendations for subsidiary stairways are given under Stage II.

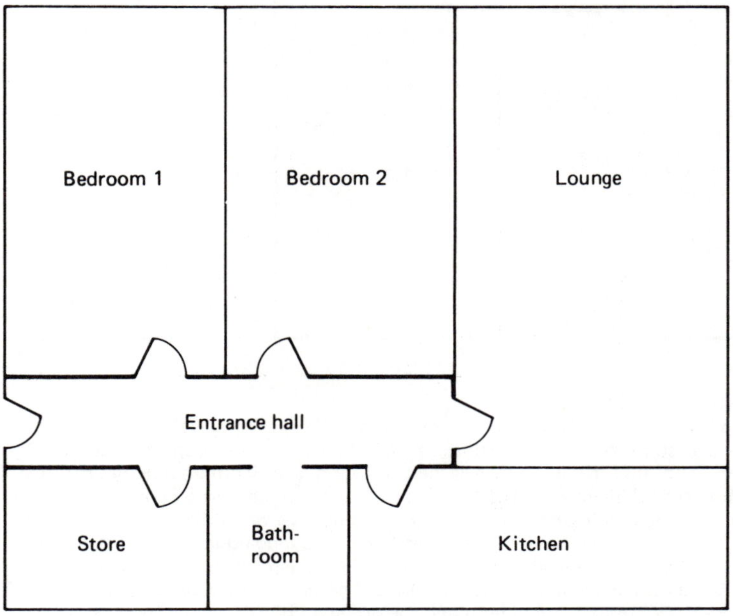

Small flat
- Walls of entrance hall – half-hour fire resistance.
- All doors except that to bathroom to be Type 3, fire-resisting.

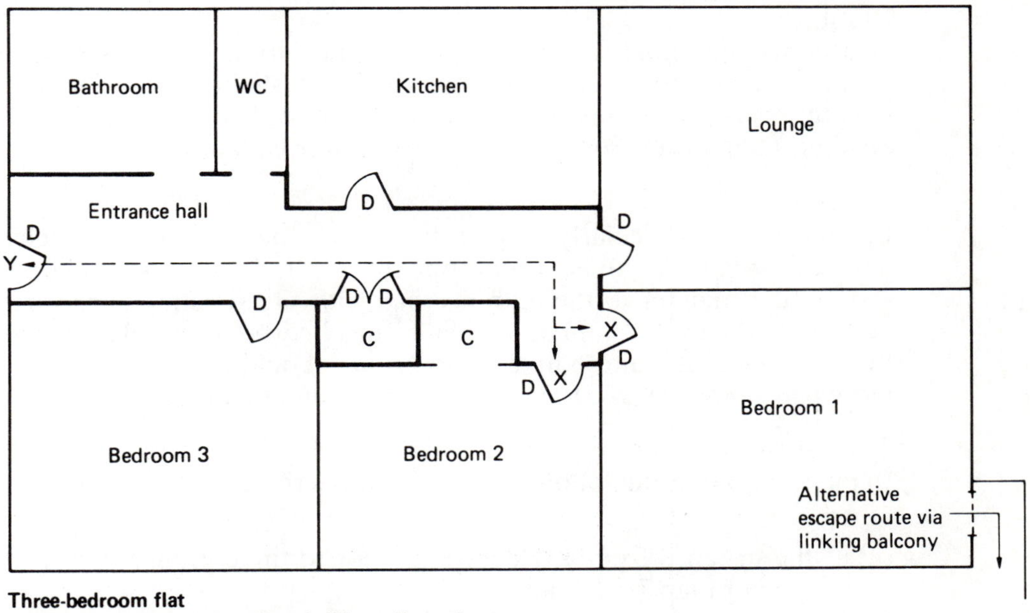

Three-bedroom flat
- Walls of entrance hall to have half-hour fire resistance.
- Doors marked "D" to be Type 3, fire-resisting.
- An alternative escape route is provided for Bedroom 1 because the distance X–Y exceeds 7.5 m and the route passes the kitchen door.
- If X–X does not exceed 1.5 m, the alternative route may also serve Bedroom 2; otherwise a pass door between Bedroom 1 and Bedroom 2 will be necessary.

FIG.2.2 Escape routes, flats with entrance halls

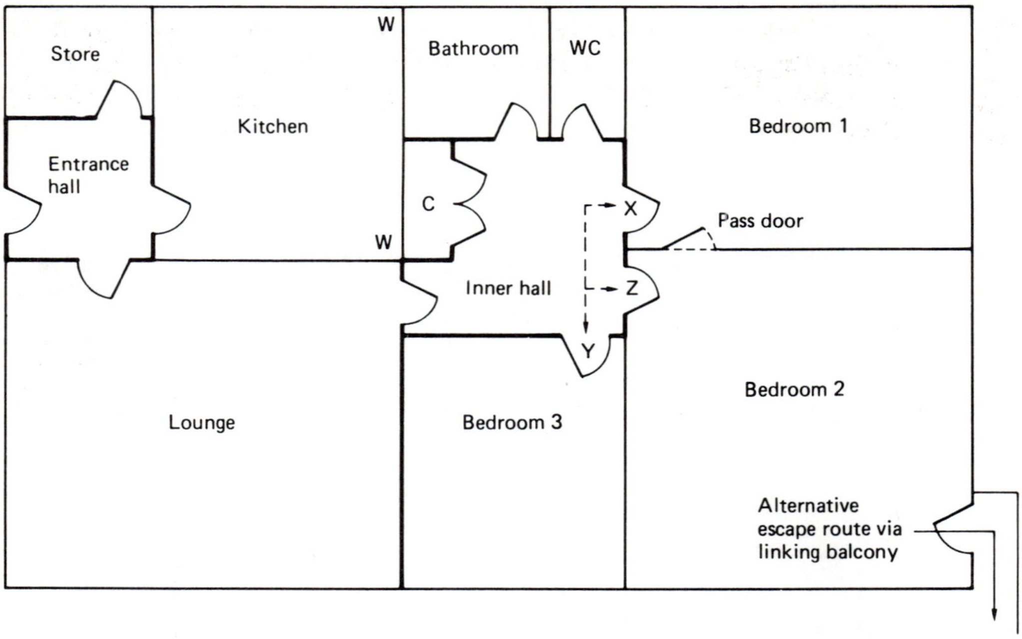

Flat with inner hall
- Both the entrance and inner halls are enclosed by half-hour fire-resisting walls.
- All doors leading on to the halls except bathroom and WC are self-closing and fire-resisting Type 3.
- The Code recommends that the wall between the living room and Bedroom 3 be fire-resisting. The safety of sleeping occupants would be improved still further by continuing this protection along the line W–W on the plan, thus completely separating the living from the sleeping part of the flat.
- Note that the Code requires the walls and doors forming the cupboard in the inner hall to be fire-resisting.
- The alternative escape route is shown taking the form of a linking balcony from Bedroom 2. As drawn, the distance X–Z exceeds 1.5 m and a pass door is therefore needed between Bedrooms 1 and 2. On the other hand, distance Y–Z is within the 1.5 m limit and the occupant of Bedroom 3 should be able to use the inner hall to reach Bedroom 2 and safety.

FIG.2.3 Escape routes, flat with inner hall

BEDROOMS

Bedroom doors should be located closer to the entrance door than either the living room or kitchen doors. No door from a bedroom should be further than 7.5m from the entrance door. If this maximum distance would be exceeded for any bedroom an alternative escape route must be provided from that room.

INNER HALL

In many cases, particularly in larger flats, it may be convenient to include an inner hall in the plan with all bedrooms opening off it. This hall should be constructed in a similar manner to the entrance hall with fire-resisting enclosure; any walls separating the living room or kitchen from a bedroom should also have half-hour fire resistance. An alternative escape route should be provided through one or more bedrooms to serve the entire bedroom/inner hall area.

TRAVEL DISTANCES

In planning a flat, the following distances are critical:

(a) the distance from any bedroom door to the entrance door should be not greater than 7.5m; and

(b) where an alternative escape route is planned from one room by way of a hall and through another room, the doors of those two rooms should not be more than 1.5m apart. Where it is not possible to achieve this maximum distance, a pass door should be provided in the wall between bedrooms.

How these distances should be measured is indicated in several of the Figures. The route is from centreline to centreline of each doorway and far enough from walls to allow for the bulk of a person; it is convenient and quick to follow the mid-line of a corridor but the line should not be closer to a wall than 270mm.

The Internal Planning of Maisonettes

Six basic accommodation arrangements, each requiring different provisions for means of escape, are shown diagrammatically in Fig. 2.4. The first four differ

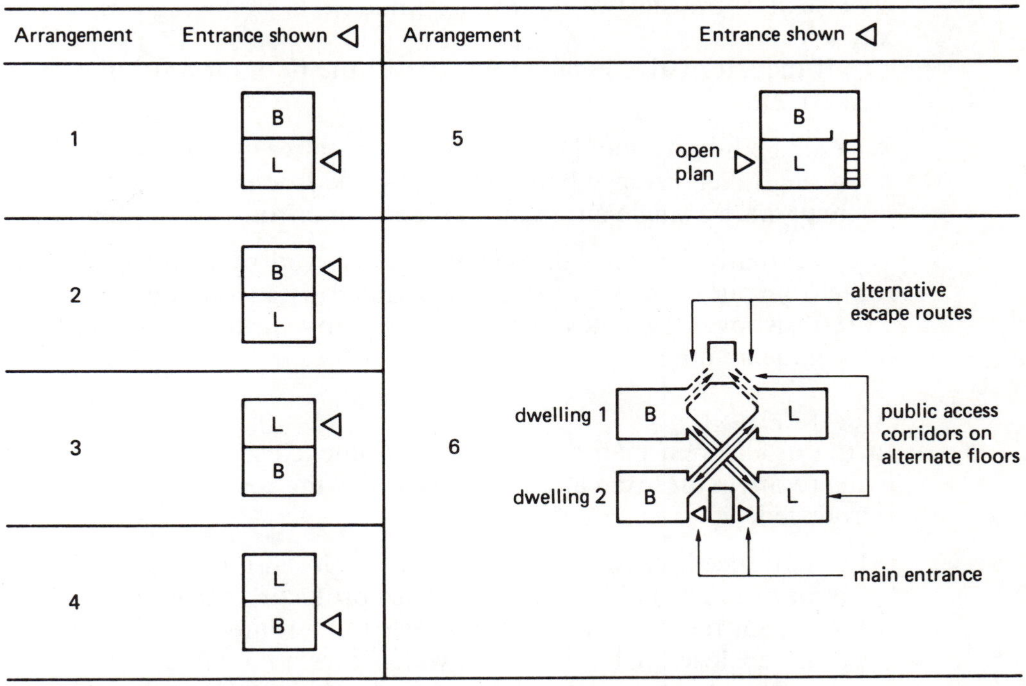

FIG.2.4 Maisonettes: examples of arrangements of accommodation

according to whether sleeping accommodation (B) is above or below the living accommodation (L) and whether the main entrance is on the upper or lower level; the fifth arrangement has an open plan living floor and unenclosed stairway. Arrangement 6 shows a more complex crossover or scissor arrangement in which each dwelling occupies space on both sides of central corridors.

Arrangements 1 to 4

There should be a circulation space within the dwelling consisting of an entrance hall, stairway and landing in which nothing should be located which could constitute a fire risk and which –

(a) is enclosed in half-hour fire-resisting construction;
(b) has self-closing, fire-resisting, Type 3 doors to living room, bedroom, kitchen and any room used for storage; and
(c) has similar doors, which need not be self-closing, to any cupboards provided for the storage of coats, linen, etc.

The living room and all bedrooms should open directly on to the circulation space; that is, none should be planned as an inner room.

The doors of the living room and kitchen should be situated as far away from the stairway as is practicable.

Whichever floor does not have the entrance door on it should be provided with an alternative means of escape providing a second exit from the dwelling which leads either –

- directly to a main escape route outside the dwelling; or
- by way of a door or low silled window, via a linking balcony, through an adjoining dwelling and so to the main escape route; or
- by way of an internal private stairway to the main escape route on another floor.

If an internal private stairway is provided as an alternative route, leading to the floor above or below the dwelling, a Type 3 fire-resisting door should be located between the two floor levels.

If the alternative escape route serves the floor containing the living room, the room exit at the start of the route should be in the living room rather than the kitchen.

If the alternative escape route serves the floor on which the bedroom(s) are located, either –

- each bedroom should have its own alternative escape route; or
- if the travel distance between any two bedroom doors is less than 1.5m, they may share a single alternative escape route from one of them.

Alternatively, all bedrooms may be served by a single alternative escape route from one of them, irrespective of travel distance along the landing, provided a fire-resisting screen and door are placed across the top of the stairway to separate it from the landing.

ARRANGEMENT 5

In the usual open plan arrangement an unenclosed stairway ascends from the living room; in these circumstances the following additional precautions should be introduced:

(a) at entrance level there should be an enclosed lobby separating the entrance door from any habitable room and so protecting Stage II of the escape route;
(b) at the entrance level an alternative escape route should be provided for any other enclosed habitable room which does not open directly on to the lobby; this should be by way of an external balcony. The balcony should be quite independent of the normal Stage II route and should lead directly to a main stairway;
(c) on the bedroom level a similar, independent balcony escape route should be provided.

Note that if all habitable rooms in the dwelling have an alternative escape route which goes by way of an external balcony to the main stairway, and all other dwellings which share the same Stage II corridor also have alternative access to the main stair by way of the balcony, the enclosed lobby at entrance level may be omitted.

ARRANGEMENT 6

The same principles as are outlined above generally apply to crossover arrangements. An escape route from the bedroom level should be provided which does not pass through the living room/kitchen level. Whichever level does not have the main entrance door should have an alternative escape route.

A typical crossover scheme has access corridors (which provide Stage II of the escape route) on alternate storeys of the building. One internal private stairway is used to connect the two levels of the dwelling; another internal stairway serves as an alternative escape route to the next higher access corridor (see diagram).

Fire doors (and screens if necessary) are so placed that smoke from a fire on the lower floor is prevented from reaching the upper one; this serves to protect the public access corridor to which the stairway which provides the alternative escape route leads.

THE ESCAPE ROUTE: STAGE II – THE COMMON HORIZONTAL ESCAPE ROUTE

Design Objective

The design objective in Stage II should be to provide one or more horizontal escape routes from every dwelling to a main stairway such that the occupants of the dwelling can reach safety without undue hazard.

The hazards which can threaten Stage II are firstly, in the case of dwellings approached by balcony or deck, heat and flame emerging from a dwelling on fire making it impossible for the occupants of other dwellings to pass the fire and reach a main stairway; and secondly, where dwellings are approached by a corridor, smoke filling part of the escape route making breathing difficult or obscuring vision for occupants wishing to escape.

The ideal, for which designers should aim, is the provision of two wholly independent routes from every dwelling to a main stairway, so arranged that the occupants can escape by moving away from the dwelling on fire without panic and without having to pass through any hazardous area. This ideal cannot always be achieved and some occupants may have only one route by which they can escape to a main stairway. This route needs additional protection to make it free of undue hazard so that it can be used at all times.

In addition, there are features which it is necessary to incorporate in this Stage to safeguard Stage III, particularly from the ingress of smoke.

Arrangements for Access and Escape

Means of escape requirements for Stage II will vary depending on such features as the arrangements for access, the number of stairways, whether the dwelling has one or two storeys and the internal planning of the dwellings. The Code distinguishes between –

- flats and maisonettes;
- rooms and dwellings which have an alternative means of escape and those from which there is a single exit;
- balcony or deck approach and corridor approach. Dwellings reached by subsidiary stairways from a balcony or deck;
- cases where a single corridor or balcony provides the sole means of escape from those where an alternative route is available. A distinction is made between alternatives which lead to a single stairway and routes which lead to different main stairways;
- single as against two stairway blocks. Point blocks need special consideration. Blocks of four storeys or less enjoy relaxations.

General Recommendations for Stage II

1. The hazard of *heat and flame emerging* from a dwelling on fire:

Dwellings approached by internal corridor will have been constructed with a fire-resisting wall and door flanking the corridor and there will be no window in the

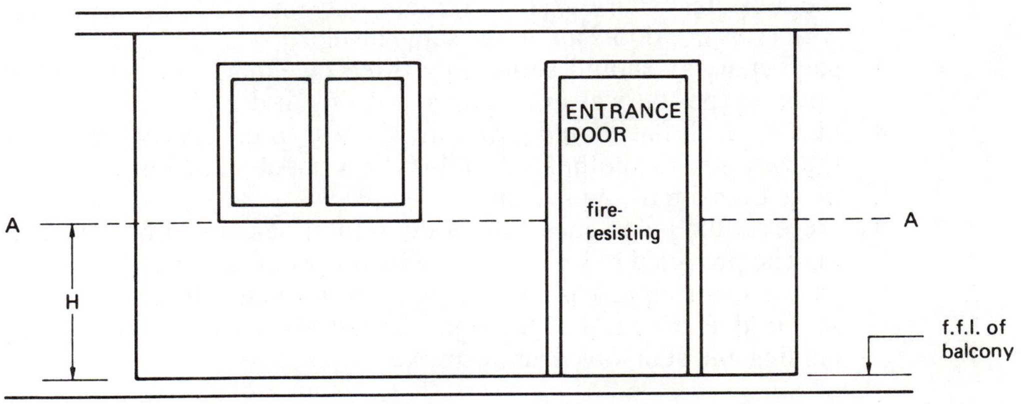

Balcony approach flat
Wall below A–A to have half-hour fire resistance and no glazing. Height H = 1.1 m min.; Door, Type 3, self-closing.

FIG.2.5 Requirement of fire resistance in walls facing onto balconies or decks

wall. This is not the case with dwellings flanking a balcony or deck. If the balcony or deck provides the sole means of escape for the occupants of any dwelling, the Code recommends fire-resisting construction for walls separating dwellings from the balcony or deck up to a height of 1.1m above finished floor level of the balcony to allow those escaping to crawl past the dwelling on fire. The requirements are illustrated in Fig. 2.5.

Some blocks with balcony approach may have balconies at the front and rear to provide alternative routes to a single stairway. Note, however that if any dwelling has a room which extends from front to rear with a window or door to both balconies, protection will be required as shown in Fig. 2.5 for either the front or the rear elevation.

2. The hazard of *smoke-logging* of corridor escape routes.

Where dwellings have balcony approach, smoke spreading into the escape route presents little problem as it will quickly disperse in the open air.

Where internal corridors provide the means of access and escape, provision must be made for dealing with smoke which penetrates the route. Two methods are recommended in the Code with variations on each. The recommendations assume that occupants will use a corridor which is only lightly smoke-logged and will also cross a short section that has become quite heavily contaminated. Maximum travel distances are specified; these are shorter in "dead-end" areas from which escape is possible in only one direction. The two alternative smoke control methods are (a) containment and (b) dispersal.

Containment: With this method smoke is contained between sets of doors positioned at intervals across the corridor; the Code makes recommendations as to type and position of door, in this way limiting the distance which any occupant may need to travel through smoke.

Dispersal: With this method, natural or mechanical ventilation is used to disperse smoke. The Code advises on the placing of permanent ventilators or openable windows (manually or automatically opened), and on the circumstances when mechanical ventilation should be provided. To ensure that through ventilation is achieved, there should be no doors across the corridor.

Subsidiary Stairways

It is possible to reduce the number of access balconies on a block of flats by arranging for the flats on some storeys to be reached by subsidiary stairways from the floor above or below.

No special measures are called for if all flats served by a subsidiary stairway have an alternative means of escape; for example, to another balcony at or above dwelling level. Where there is no such alternative means of escape the Code recommends that –

- the subsidiary stairway should extend only one storey up or down to serve flats with entrance doors all on the same level;
- each stairway should serve only three dwellings: no other dwelling should open off the stairway enclosure at balcony or deck level;
- ideally, each flat should have direct access to the stair without having to pass the entrance to another flat; if this is not possible, no entrance door should be more than 5m from the stair;
- there should be no enclosure of the stair at balcony or deck level but a screen may be provided to keep the area clear of snow and ice;
- where the stairway is one giving access to the floor above, ventilation is required at the head of the stair, ducted to outside as necessary, to keep the landing and stairway clear of smoke.

Detail Recommendations for Stage II

The recommendations of the Code are illustrated in a series of diagrams (Figs. 2.6 to 2.8).

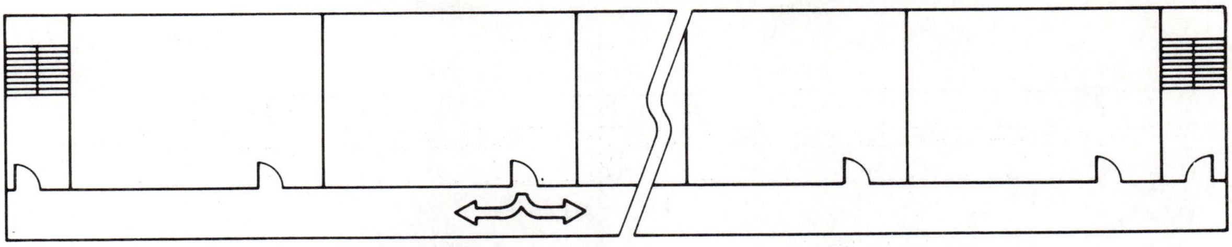

(a) Two stairways providing alternative routes from every dwelling.

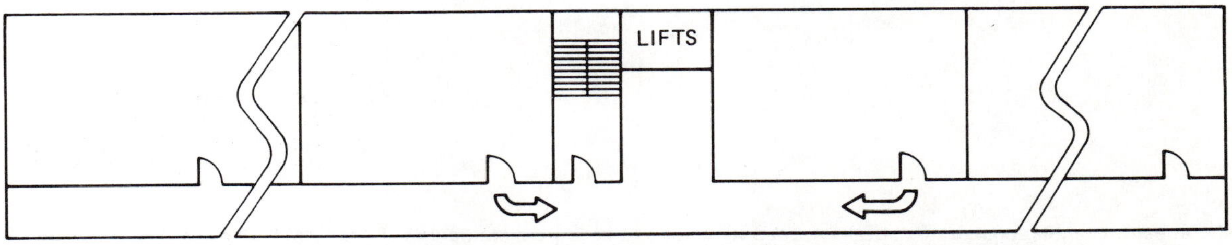

(b) Single stairway: the wall and door between the dwelling and the balcony to be protected as shown in Fig. 2.5

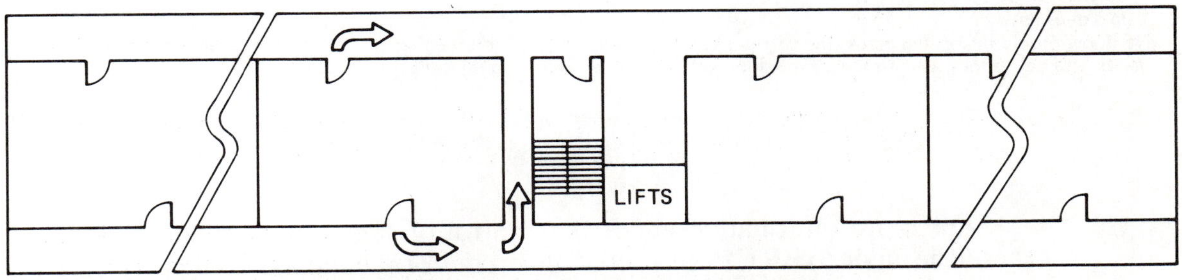

(c) Single stairway: the second balcony provides an alternative route from every dwelling.

Travel distance
The Mandatory Rules prescribe no maximum travel distance but fire-fighting requirements may impose a limit in any of these arrangements.

FIG.2.6 Dwellings with balcony access: flats or maisonettes

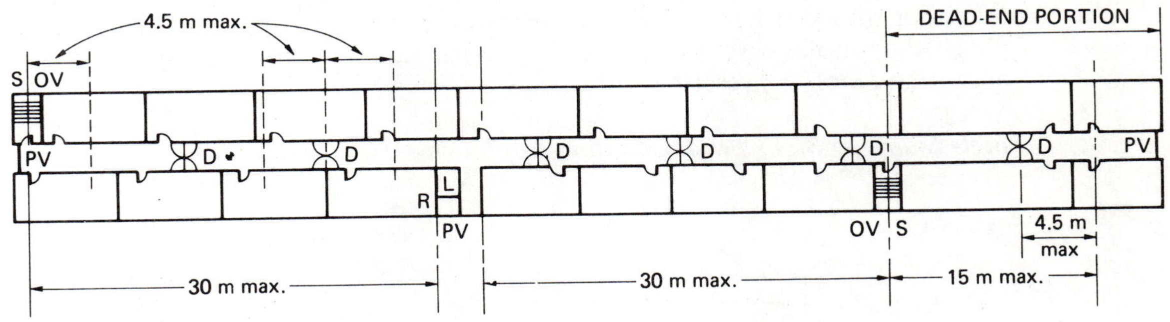

Key

S = main stairway
L = lifts
R = refuse chute
PV = permanent ventilation
OV = openable ventilation
D = Type 4, double-swing fire doors

Notes

1. No entrance door to a dwelling to be more than 30 m (15 m in a "dead-end") from either a door on to a main stairway, or a doorway to a stairway protection lobby.
2. The entrance door to a dwelling shall be not more than 4.5 m from either a stairway door, or a Type 4 door across the corridor. This rule determines the location of doors (but see Note 3) which ensure the containment of smoke and limit the distance anyone may have to travel through a smoke-logged area.
3. If any dwelling is provided with an alternative means of escape through an adjoining dwelling by way of a linking balcony, a Type 4 door should be located across the corridor between the entrance doors of the two dwellings.
4. Although not a requirement of the Mandatory Rules, the Fire Authority may stipulate that one lift be designated a "fire lift" and that it shall be located within 15 m of a main stairway.

FIG.2.7 Corridor approach – smoke containment. (The majority of dwellings have access to two stairways, but this illustration also depicts those which are located in a "dead-end".)

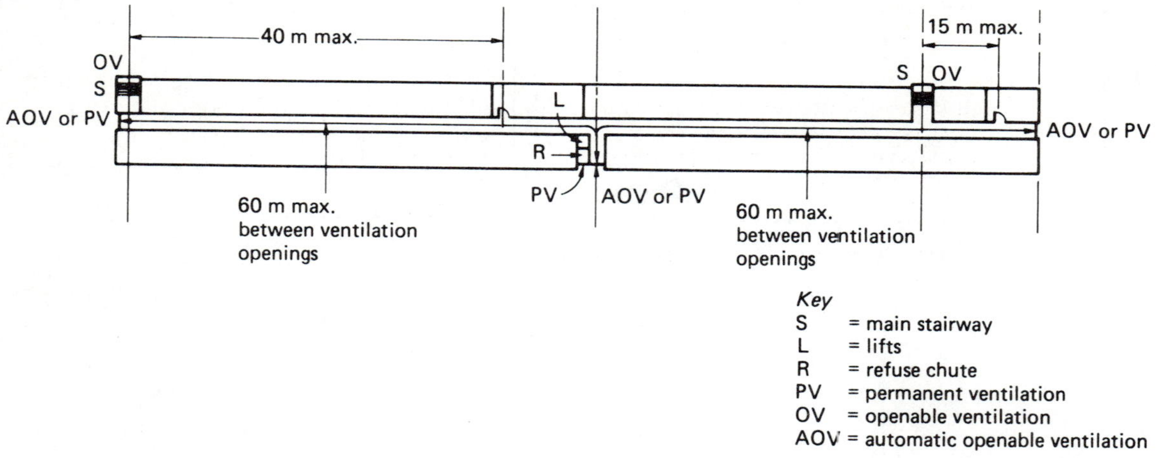

Notes

1. Permanent ventilation or automatically opened ventilation to be located at intervals of nor more than 60 m. The automatic ventilation should be controlled by smoke detectors fitted in corridor ceilings at not more than 20 m centres.
2. The distance between the entrance door of a dwelling and the entrance to a main stairway must not exceed –
   between two stairways 40 m
   in a dead-end 15 m

   No doors should be located across the corridor as they would prevent cross-ventilation. This applies even when a dwelling has an alternative means of escape via a linking balcony and an adjoining dwelling.

FIG.2.8 Corridor approach – smoke dispersal. (This illustration, like Fig.2.7, includes the case of a dwelling in a corridor "dead-end" with access to one stairway only.)

For more information about the control of smoke in corridors, reference should be made to BRE Digest 260, *Smoke control in buildings: design principles*.

In all cases of balcony or deck approach to dwellings, the Mandatory Rules for Means of Escape impose no limits on Stage II travel distances between a flat entrance door (or the point where a subsidiary stair meets the balcony) and a main stairway. However, the fire service may impose limits for fire fighting reasons. Where the escape route follows a corridor and the smoke containment method is adopted, there is a prescribed maximum distance between the door of a dwelling and either the door into a stairway enclosure or a fire door across the corridor. Where the smoke dispersal method is used, the maximum distance between ventilation points is prescribed.

Single Stairway Blocks with Corridor Approach

Note
The Code does not show balconies across the ends. The balconies would seem to be necessary if a safe, alternative route to the stairway is to be provided for dwellings on the non-stairway side of the corridor.

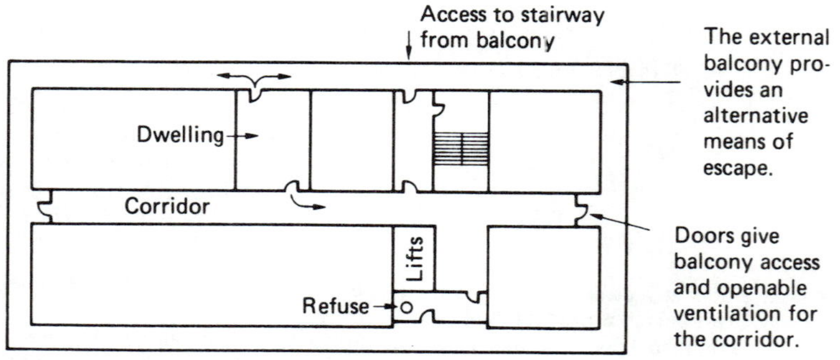

FIG.2.9 Single-stairway buildings and the provision of balconies

Every dwelling has an alternative escape route (wholly independent of the normal approach route) which is either –

- by balcony* giving access to the stairway; or

- by a private stairway† leading to a common access corridor on another floor, preferably the one below.

* If unconnected balconies along each side of the building are used, protection as detailed in Fig. 2.5 will be necessary.

† May serve more than one dwelling, provided entry from each dwelling is independent and a Type 3 door is located between the stairway and the access corridor.

The distance between the door into the stairway enclosure (or stairway protection lobby) and the entrance door of a dwelling (or door from a private stairway) shall not exceed 40m.

Requirements for Ventilation of Common Access Corridor

- If the corridor exceeds 30m in length, it must be extended to the external wall at both ends.
- If the corridor is 30m or less in length ventilation at one end is sufficient.
- Ventilation should be provided by means of a window or door giving a free area of at least 1.5m^2. Or the corridor can be ventilated indirectly through a lobby.

THE ESCAPE ROUTE: STAGE III – MAIN STAIRWAYS: THE COMMON VERTICAL ESCAPE ROUTE

General Considerations

In referring to stairways which are used for escape the Code uses two terms, "main stairway" and "protected stairway". Both terms refer to a stairway which provides Stage III of the escape route. It is not possible to have a main stairway which is not also protected.

Building Regulations require all stairway enclosures to be constructed as "protected shafts"; detailed requirements are given in the Approved Document B2/3/4, para. 2.30 and Appendix E; Table 2.1 prescribes fire resistance periods for the enclosing structure. Where there are four or more storeys, materials used for the construction of walls, stairs and landings must be of "limited combustibility" as defined in Appendix A of the Approved Document. The protection of openings is dealt with in Appendix F.

The enclosure of a stairway must be continuous for its full height except where, at its lower end, it discharges to the ground (or an open podium) via a permanently open flight, or where at the top it extends above the general roof level (for example, to give access to the roof).

With these exceptions, Stage III of the escape route comprises the enclosed stairway and, in addition, the horizontal connection from the foot of the stairway to the final exit. This hall or corridor must also be enclosed in fire-resisting construction and there are constraints on ancillary accommodation which can be located within the enclosure. Lifts are never accepted as a satisfactory means of escape from fire. (But see notes dealing with facilities for the disabled.) The Code describes the circumstances when a main stairway may, with suitable precautions, be continued down to basement level.

A main stairway is entered at each storey through a self-closing fire door. When the route between the stairway and the dwellings is an open external balcony, the single door provides adequate protection for the stairway; but with corridor approach additional safeguards may be needed. These take the form either of smoke control doors across the corridor at prescribed intervals to contain smoke, or corridor ventilation to disperse smoke, or permanently ventilated stairway protection lobbies. The proper solution in particular circumstances may be a combination of these features. The case of the single staircase building is one which needs the most careful planning (see p. 52).

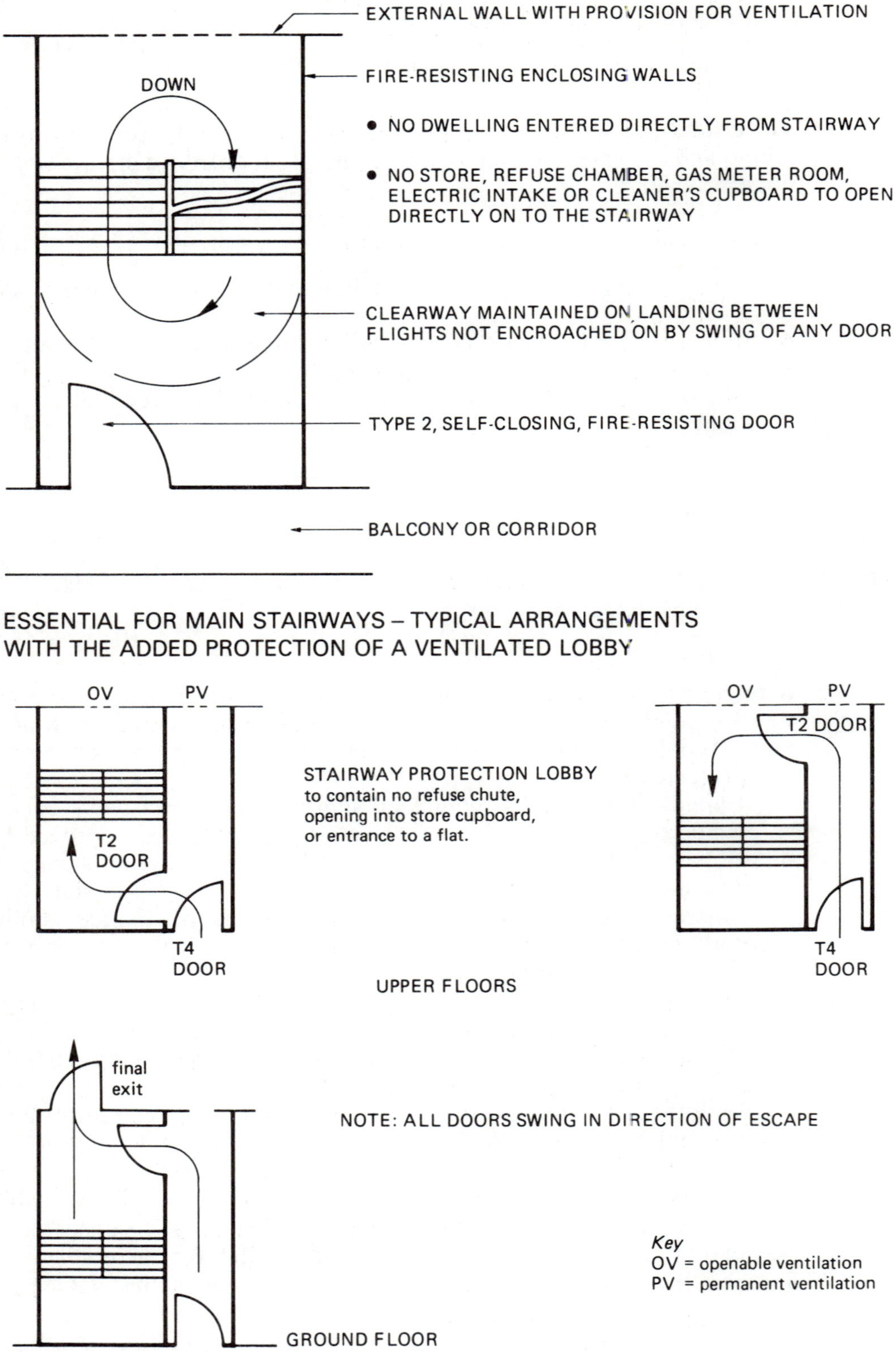

FIG.2.10 Stairway protection

The main layout requirements for stairway enclosures and the arrangement of their associated protection lobbies are illustrated in Fig. 2.10. In addition the Code recommends the following dimensional limits for main stairways forming part of an escape route.

Minimum width:	1000mm. Where a handrail projects from the wall the width may be reduced to 900mm.
Treads:	240mm min. (measured from nosing to nosing).
Risers:	190mm max. Not less than 2 nor more than 16 risers in one flight.
Pitch:	max. angle 38°.
Headroom:	nowhere less than 2m measured vertically above the stair nosings.

Flats may be approached by a single stairway in a multi-storey building provided certain safeguards are incorporated to protect the stairway. The figure shows four possible routes (marked A, B, C and D) between a dwelling and the stairway with features introduced along the route to isolate the stairway from the effects of fire in a dwelling. Each starts from a dwelling; A is appropriate for blocks served by an external balcony or deck: B, C and D show alternative arrangements recommended by the Code for dwellings approached through internal corridors and/or lobbies.

Travel distance restrictions apply in some zones: 15 m is the limit in a smoke dispersal corridor (example B) and in a smoke containment corridor (example C); 4.5 m applies in a common hall or lobby (example D). The permitted location of lifts is indicated in dotted outline.

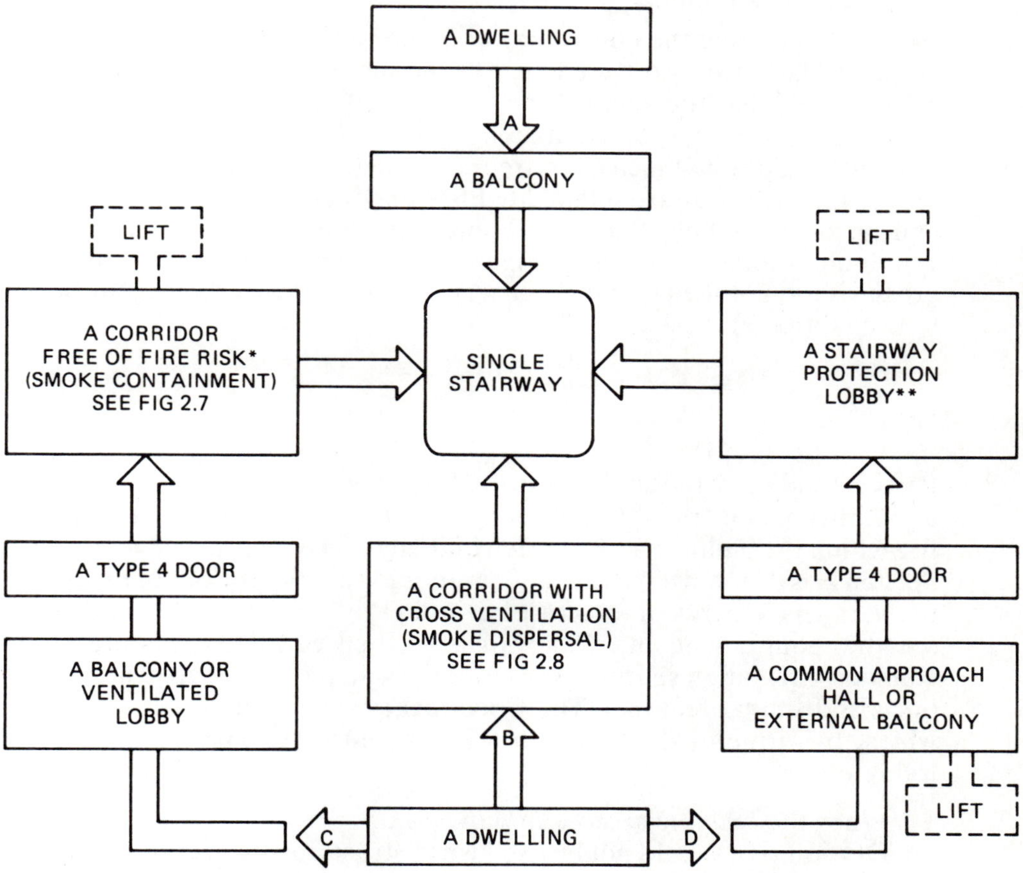

*Into which no dwelling, refuse chamber, store, etc., opens

**Which

- adjoins an external wall,
- is permanently ventilated
- is free of fire risk*
- communicates only with
  - the stairway;
  - a lift; or
  - an external balcony

Note less onerous requirements apply if the building is less than four storeys high.

FIG.2.11 Requirements for single-stairway buildings

These dimensions correspond with those recommended in Approved Document K for common stairways. Note that the minimum width should be maintained on landings to provide a clear path between flights (see Fig. 2.9). Doorways opening onto the stairway from balconies, corridors or lobbies should be located so that the swing of the door does not encroach on the clear path to impede the passage of people using the stairs.

Main stairways should not contain any fire hazard. Service ducts for gas or electricity should not be run within the enclosure and no dwelling, store or refuse chute should open directly on to a stairway.

Balustrading, a screen or railings are required to guard any situation from which a person using the stairway in an emergency might fall. At least one side of every stairway should be provided with a handrail, which should be continuous (except that it may end two steps short of the foot of the stairway). If the width of the stairway exceeds 1000mm handrails are required on both sides. Normally, handrail height should be between 840mm and 1000mm, but if the handrail forms the top of a guarding 900mm is the minimum on flights and 1000mm on landings.

There are also requirements for emergency lighting on stairways. Protected circuits, independent of circuits elsewhere in the building, are required and there must be provision for over-riding any timed switching.

Single Stairway Buildings

Because it provides the only route out of the building, the single stairway needs to be particularly well protected against the ingress of fire and smoke. The Code recommends that the stairway be approached either from a balcony or from a ventilated "stairway protection lobby" (see Definitions). In the case of the lobby approach, additional measures are recommended for the corridor leading to the lobby; the measures are intended either to contain smoke within a limited zone and make it unlikely that it will spread to the stairway, or to disperse it by ventilation to outside air. Figure 2.11 summarises the options. Slight pressurisation of the air in the stairway is another device which may be employed to keep it free of smoke.

CONSTRUCTION

Parts B2 and B3 of the Schedule to the Regulations require materials to be selected and forms of construction to be adopted for buildings so as to limit the spread of fire within the building and between buildings. The Approved Document, B2/3/4, Section 2, is devoted to "Flats" and prescribes suitable periods of fire resistance for the various elements of structure according to the size of the building. It makes specific requirements in relation to walls (between flats and between flats and common circulation spaces), walls forming stairway enclosures, lift shafts, service ducts and similar features. The Code makes some additional recommendations which, by virtue of the Mandatory Rules, must be complied with. They are as follows:

1. When the *building exceeds 15m in height*:
   Dry-jointing should not be used when structural connections are being made. Wall to wall and wall to floor junctions should be monolithic and incorporate durable features for the transfer of load between elements; they should be protected from the effects of fire and should prevent fire penetration for the same period as is required for the elements joined.

   Continuous membranes should not be used to provide the fire protection for several elements; rather, elements should be individually encased. Approved Document B2/3/4 requires cavities behind or within such protection to be fire-stopped.

2. In *all blocks of flats* irrespective of height:
   Attention must be given to the junctions between elements to prevent fire

Table 2.1 Summary specification for fire-resisting doors for use to protect escape routes

DOOR TYPE† (CP3)	FIRE-RESISTANCE PERIOD (minutes)		SELF-CLOSING DEVICE REQUIRED	MINIMUM REBATE Depth to frame		NUMBER OF LEAVES permitted	REBATE REQUIRED at meeting stiles?	SWING of leaf or each leaf	USE The principal locations and/or uses of door types
	Stability	Integrity		Under test	In use				
1	as prescribed period for the wall		yes, NOT rising-butt hinges	25mm	25mm	1	N.A.	one direction only	to delay fire spread into escape routes from plant rooms, boiler rooms, basements, garages and service ducts.
2	30	30	yes, NOT rising-butt hinges	25mm	none*, or un-specified depth	1 or 2	No*	one or both directions	to limit smoke spread into main stairways from corridors and ventilated lobbies.
3	30	20	yes, may be by rising-butt hinges	25mm	12mm	1 or 2 if with opposing swings	yes, 12mm min	one, or two if in opposite directions	within dwellings, between a habitable room and a hall or landing; also between a dwelling and a common circulation area.
4	30	20	yes, NOT rising butt hinges	25mm	may be none*	1 or 2	no*	one or both directions	for smoke control (containment) in corridors and between corridors and ventilated lobbies.

† Type 1, 2 and 4 doors should carry a warning notice that they are provided for fire safety and should be kept closed. They should not be provided with cabin hooks or similar fitment.

* Where doors are fitted in frames without rebates, or when meeting stiles are not rebated, clearances between door and frame or leaf and leaf should be as small as is reasonably practicable but not greater than 3mm.

finding a way through from one compartment to another. The Code refers specifically to lightweight, non-loadbearing partitions or screens. These frequently need to have a movement joint around the perimeter to accommodate relative movements; to be effective the joint must be left open; it may however be masked by a cover strip fixed to one side. This results in a line of weakness through which fire may penetrate. It is a requirement that such joints be not more than 13mm wide and that they be tightly packed with a mineral fibre material.

FIRE-RESISTING AND FIRE-CHECK DOORS

The British Standard, BS 476: Part 8: 1972 specifies the test procedure for fire-resisting doors. Performance is judged by reference to three criteria. They are:

"stability":	that is, resistance to collapse;
"integrity":	that is, resistance to penetration by flame or hot gases and to spontaneous flaming on the unexposed face; and
"insulation":	that is, resistance to the passage of heat; so that the temperature rise on the unexposed face is limited.

Building Regulations and Codes, including CP 3: Chap. IV: Part 1, usually require compliance with only the first two criteria.

A fire-resisting door tested in accordance with BS 476 is graded according to the period for which it satisfies the relevant criteria. Thus a door may be graded as a "half-hour door" or a "one-hour door". However, when the main function of a door is to arrest the passage of smoke, a modified grading for "integrity" is acceptable. In this case the door may be identified as a 30/20 door, indicating that the door satisfied the stability criterion for 30 minutes and the integrity criterion for 20 minutes. In the case of fire doors it is important to realise that the entire assembly of door, door frame with rebated stops and door furniture (hinges, latches, etc.) contributes to its fire grading.

The most recent recommendations made for fire-resisting doors for use to protect escape routes recognise that "integrity" is the key criterion and the designer may find manufacturers of doors specifying the product as an FD 30 or an FD 20 door; this indicates a door which has satisfied the integrity criterion of the fire test for a period of minutes indicated by the numeral.

The Code for Means of Escape from flats specifies four door "Types" for use in different situations; the specification makes use of the fire-resistance grading described above. The fire-resistance requirement and other characteristics are summarised in Table A.

Another British Standard, BS 459: Pt 3: 1951, supplemented by various Published Documents (PD) and Amendments (AMD), specifies materials and construction of "Fire-check flush doors and wood and metal frames (half-hour and one-hour types)". Useful information can also be found in BRE Digest 220, *Timber fire doors*.

VENTILATION

Provision of ventilation to remove smoke is required for –

(a) a stairway enclosure;
(b) a stairway protection lobby;
(c) a corridor;
(d) a ventilated lobby; or
(e) a compartment in which a refuse chute is located.

A subsidiary stairway serving dwellings on the floor above the access floor is also required to have ventilation at its head.

Each of these zones is required to be ventilated directly to outside air; for example, a corridor should not be ventilated through a stairway protection lobby. However, the Code does describe particular circumstances when it is permissible to ventilate a corridor through a stairway enclosure.

Ventilation requirements are shown in the illustrations by the following letters which have the meanings indicated:

OV = a ventilator which is openable by hand. It is mainly for use in the later stages of a fire and it will generally be a member of the fire brigade who will open the ventilators.

PV = a ventilator which is permanently open to outside air; by careful positioning or the use of louvres, penetration of rain or snow can be limited.

AOV = a ventilator which is automatically opened by linking it to smoke detectors in the area to be ventilated.

Permanent and automatic ventilators clear smoke from the earliest stages of a fire. In prescribing requirements it is acceptable in some circumstances to have part of the area at each ventilation position in the form of PV or AOV, the remainder being available as OV if needed.

When the zone to be ventilated does not adjoin an external wall, ducting will be required.

For natural ventilation to perform satisfactorily, a cross-flow of air must be achieved; thus, in corridors, ventilators need to be positioned at both ends and in a long corridor intermediate connection to outside will be needed. In cross-ventilated corridors the Code recommends a maximum distance of 60m between direct connections to outside walls. Ventilation of stairways is often best achieved by a combination of windows at each storey level coupled with some open area at the top to induce a stack effect.

The area of ventilator needed to keep corridors acceptably free from smoke when the smoke dispersal method is employed depends on the location of the ventilators. Zones of pressure and suction occur on different faces of the building according to wind direction and speed; to achieve good through ventilation, openings to outside air should be located on different elevations. BS 5925 is the Code of Practice for natural ventilation and will be of assistance in deciding ventilator areas and locations; BRE Digest 141, *Wind environment around buildings*, also contains relevant information.

The ventilator area and spacing requirements prescribed in the Code assume a minimum wind speed of 5.5kph, satisfactory distribution of ventilation openings and good internal discipline regarding the opening and closing of doors and windows. That accepted, the Code requires that each ventilator should provide a minimum free area of opening of 1.5m^2. In a corridor where the smoke dispersal method is adopted, one-third of the ventilation area required at any position (0.5m^2) should be PV or AOV; the remainder may be OV.

Where a main stairway is not on an external wall so that it cannot be provided with openable windows, permanent ventilation of at least 1.0m^2 free area via a duct is required.

Any system of ducted air heating, ducted ventilation or air conditioning should be so designed that smoke or fire cannot be carried through the system to places where it could jeopardise escape.

Stages II and III of the escape route may, in certain cases, be kept free of smoke by the use of a pressurisation system of sufficient power to counteract the pressures generated by combustion and the buoyancy effects of hot gases.

ANCILLARY ACCOMMODATION

The facilities and accommodation listed below represent potential fire hazards and should not open directly on to an escape routes which provides the only escape route from a dwelling. Thus, if every dwelling has an alternative route in Stage II, but both lead to a single stairway, a porter's room should not open directly off the ground floor hallway which connects the stairway to the final exit.

RECEPTION AREA, PORTER'S OFFICE OR SIMILAR

In a building with two or more stairways, a reception area (with desk and chairs) may be located in one of the hallways. Anything more elaborate, like an office, must be separated from the hallway by a fire resisting wall or screen with fire doors giving access through a lobby.

CUPBOARD, STORE ROOM

These include drying rooms, pram stores, cupboards for cleaning materials and rooms housing service intakes with or without meters. These should only open directly off a part of an escape route when an alternative route is available to bypass that point. They should never be located on the sole escape route. In all cases they should have fire resisting walls and floors of one hour minimum period and they should be entered through a self-closing, Type 1, half-hour door. The door should be kept locked and marked "KEEP LOCKED".

TRANSFORMER ROOM SUB-STATION

These chambers should not be accessible from within the building. Located on an external wall, enclosing structure should have not less than two hours fire resistance in non-combustible construction. If oil immersed gear is involved, an oil retaining sill should be placed across the door threshold.

CENTRAL BOILER ROOM

Where boiler rooms are located at the bottom of the building the Code makes the following recommendations:

1 As a first preference:
   - complete separation of the boiler room from the remainder of the building by imperforate walls and floor providing 2hr fire resistance;
   - access to the boiler room only from outside the building;
   - the door, or other access, to be located well away from the final exit of any escape route.
2 If all of the above recommendations cannot be complied with and internal communication is necessary:
   - enclosing walls and floor to have 2hr fire resistance and the door leading to the building to be a 2hr Type 1 door with automatic self-closing by fusible link;
   - when not in use the door must be locked; it should be marked "KEEP LOCKED".
   - the door must not open into storage accommodation;
   - the door must not be located where smoke or fire discharging from it might obstruct an escape route;
   - adequate provision should be made for the supply of combustion air to the boilers directly from outside. It should never be necessary to open the door between the boiler room and the building to augment the air supply. The air supply provision should anticipate future changes of fuel.

Where an oil-fired boiler is to be located at the top of the building, the requirements are as follows:

- enclosing walls and floor to have 2hr fire resistance and similar provisions apply to the door;
- the main oil storage tank must be located at ground level with a service tank (capacity not to exceed 910 litre) at boiler room level but in a separate chamber;
- the service tank chamber to be totally enclosed by non-combustible construction providing 4hr fire resistance;

- the oil supply pipe between main and service tanks to be enclosed in its own duct having 4hr fire resistance;
- arrangement to be made to drain down the service tank automatically in the event of fire in the boiler room.

CAR PARKS AND GARAGES

If it is intended to locate a car park or garage greater than 465m^2 in area, or to store petrol under or adjacent to a block of flats, the Petroleum (Regulation) Acts 1928 and 1936 require the local authority to be consulted and a licence to be issued by them.

For a smaller garage having no petrol storage a licence is not necessary. In this case the following provisions apply:

- the garage should be separated from the main building by structure providing 1hr fire resistance;
- if the building has a single stairway, there must be no internal connection between garage and building;
- however, if internal connection is essential, a 1hr fire-resisting door (Type 1) must be located so as to prevent fire spreading from the garage into the building. The connection should only be linked with a stairway through a stairway protection lobby.

SERVICES

Stairway and Corridor Lighting

Stages II and III of the escape route should be adequately lit at all times. The lighting must remain on despite any damage to circuits caused by a fire in a dwelling. Corridors, stairways and the entrance hall should be supplied by independent protected circuits; all wiring should be run along routes of negligible fire risk. The stairway circuit should be connected directly to the main switchboard; if a stairway is an internal one and has no natural lighting the circuit should be independent of the corridor lighting.

It is a common practice to have some form of automatic time switching for stairways. If this is provided, there should be a switch on each landing which can be used to override the automatic system and switch lights on permanently; this switch should be incapable of switching the lights off again.

Gas and Electricity Services

The Approved Document (B2/3/4) contains requirements for service ducts. Firstly, they must be fire-resisting; the Code recommends a minimum period of fire resistance of half an hour, but Regulations require one hour if the building has three storeys or more. The materials of construction have to be of limited combustibility and the exposed internal surfaces should be chosen so as not to assist fire spread. Fire-stopping is required around the outside of the duct wherever it passes through a compartment wall or floor.

Codes of Practice are available which deal with the safe installation of most service systems. Building Regulations in Part J make requirements for the structure associated with heat-producing appliances. The Code for means of escape from fire relies on the regulations, but emphasises some points of particular importance.

Gas and electricity services should not be run through the same duct. The Code reinforces the Building Regulations requirement that a duct carrying a gas service pipe must be ventilated at top and bottom and that there should be no internal fire-stopping. Doors or inspection covers giving access to any duct should have the appropriate fire-resistance.

Within dwellings, gas and electricity meters should not, if possible, be located in the entrance hall; if the hall is the only possible place for them, they should be housed separately, each in a non-combustible cupboard. Where meter boxes are used to allow reading to be done from outside the dwelling, gas and electricity should be in separate boxes. Behind the boxes the requisite fire resistance of the wall must be maintained.

Heating Systems in Dwellings

Heating systems in dwellings generally have no effect on the safety of means of escape. However, the Code draws attention to two features which could contribute to rapid fire growth. Firstly, if there are individual oil-fired boilers inside dwellings, precautions should be taken to confine any leakage of oil to a sump below the storage tank. The tank should be of strictly limited capacity. The pipe supplying the boiler should have an automatic cut-off valve or a manual valve which is easily reached.

The second feature to which attention should be given concerns any ducted warm air heating system. This should be designed so that the system does not spread smoke or fire from rooms to circulation spaces or from living rooms to bedrooms. Return air grilles at ceiling level should be avoided since this is the zone in which smoke from a fire first begins to build up.

APPENDIX

Relevant clauses of BS CP3: Chapter IV: Part 1: 1971

The following clauses of the BS Code of Practice, CP 3: Chapter IV: Part 1: 1971 are cited in para. 2.3 of Section 2 of the Mandatory Rules. They are listed here together with an indication of the subject dealt with by each clause.

SECTION 1 General
- 1.2 Definitions

SECTION 3 Planning Recommendations
- 3.1 General
- 3.2 Internal planning of flats and maisonettes
  1. Flats
  2. Flats entered from a floor above or below the flat
  3. Maisonettes
- 3.3 Stage II: Escape from flats to main stairways
  - 1 Dwellings with balcony or deck approach
  - 2 Dwellings with entrance doors opening indirectly off a balcony or deck via a subsidiary stairway
  - 3 Dwellings with corridor approach where every dwelling has an independent alternative escape route to a main stairway
  - 4 Dwellings with corridor approach where one or more dwelling has no alternative escape route to a main stairway
    - (a) protect escape route by smoke containment
    - (b) protect escape route by smoke dispersal
- 3.4 Stage III: Main stairways
  - 1 Enclosure
  - 2 Design
  - 3 Single stairway conditions
  - 4 Multi-stairway conditions
  - 5 Stairways without permanent ventilation
  - 6 Ventilation to internal stairways

SECTION 4 Construction
- 4.1 General
- 4.2 Fire protection
- 4.3 Fire resisting doors
- 4.4 Stairway enclosure

MANDATORY RULES: SECTION 3

Shops

CONTENTS

MANDATORY RULES: SECTION 3

Shops

INTRODUCTION

The Requirement

The requirement is that –

"... there shall be means of escape in case of fire from the building to a place of safety outside the building capable of being safely and effectively used at all material times."

Satisfying the Requirement

The requirement can only be satisfied by complying with the Mandatory Rules and they in turn require that the building shall be constructed in accordance with the recommendations of certain named sections of British Standard 5588: Part 2: 1985. The named sections are listed in Appendix 3.1. Throughout these notes this BS is referred to as "the Code".

Application

Paragraph B1 in Part B of Schedule 1 to the regulations, in the section headed "Limits on application" states in relation to shops –

"1 This requirement applies only to:
(a) a building which is erected and which ...
(iv) is or contains a shop;"

"2 The means of escape provided need only, ... in the case of a building containing ... a shop, afford escape for the people from the ... shop."

Appendix A of the Mandatory Rules defines a SHOP; the definition means that the Rules apply to premises used for any of the following purposes –

1 For retail trade or business, including sales by auction; these are shops in the everyday sense of the word; they range from small shops to department stores, supermarkets and hypermarkets.
2 For sale to the public of food or drink for immediate consumption: this group includes cafes, restaurants, public houses, bars, etc. The words "immediate consumption" seem to exclude take-aways.
3 For lending books or periodicals for profit.
4 For hairdressing or use as a barber's shop.
5 For the receipt of goods brought in for repair or treatment: for example, TV or shoe repair; also premises which offer the public facilities for repairing or treating goods themselves; for example, laundrettes.

In the Code it is suggested that some premises which might be thought of as shops are better treated as offices because of the nature of the activity carried on in them; the Code mentions banks and estate agents. Post offices, betting shops and similar establishments, in which few if any goods are displayed for sale and in which the activities are mostly clerical, would also seem to qualify for this category.

Paragraph B1 of Schedule 1 applies to the erection of a building which is or contains a shop. In the introductory remarks to these notes on means of escape (page 10), it was held that the Mandatory Rules do not apply to an existing building which is to be extended or materially altered to provide a shop. Reasons were given for this interpretation. However, reference was also made to a memorandum issued by the DoE which puts forward a contrary view. If the DoE view is accepted, the recommendations of the relevant clauses of the Code will have to be followed, a building notice cannot be given and full plans must be deposited. The local authority must consult the fire authority in accordance with s.16 of the Fire Precautions Act 1971.

It should be noted that, under the Fire Precautions Act, a shop is a "designated building" and, even if exempt from the requirement to have a fire certificate under the Fire Precautions (Factories, Offices, Shops and Railway Premises) Order 1989, is subject to the supervisory powers of the fire authority on means of escape and means for fighting fire in accordance with codes of practice issued by the Home Secretary.

Definitions

ACCESS ROOM Provides the only escape route from an inner room; see Fig. 3.9.

ACCOMMODATION STAIRWAY
An unprotected stairway connecting two or more shop floors; it is not part of any escape route and may be an escalator.

ANCILLARY ACCOMMODATION
Rooms or parts of the building used for offices, storage, plant, etc., which support the main activity of selling.

DEAD END An area of a shop floor or of ancillary accommodation from which only one route is available to an exit; or to two exits the routes to which do not diverge by 45° or more (see Figs. 3.2 and 3.3).

COMPARTMENT
A zone of the building separated from the remainder of the building by walls and floors (constructed in accordance with the recommendations of Approved Document B2/3/4, paras. 6.23 to 6.28, for compartment walls and floors).

DIRECT DISTANCE
See Fig. 3.1. The straight-line distance between a point on a storey and a storey exit.

FINAL EXIT The point at which an escape route discharges to a place of safety outside the building, usually at ground level.

FIRE DOOR A door or shutter provided to close an opening in any enclosing or protecting structure, or in a shaft or duct; the door, door furniture and hinges are designed to resist the passage of fire for a specified period.

HEIGHT OF BUILDING
A measurement taken from the level of the ground or pavement, at the centre of the tallest side of the building, to the highest point of the floor of the top storey. (A top storey housing only plant is excluded.)

HIGH FIRE RISK AREA
Areas which, because of the use made of them or materials stored on them, constitute a higher fire risk than a sales floor.

INNER ROOM A room with only one exit which leads into an access room (see above); see Fig. 3.9.

PRESSURISATION
A method of reducing the penetration of smoke into protected zones by maintaining them at a raised air pressure.

PROTECTED CORRIDOR, LOBBY OR STAIRWAY
A part of an escape route which is separated from the remainder of the building by fire resisting construction.

STOREY EXIT A doorway between unprotected and protected zones of the escape route.

TRAVEL DISTANCE
The distance which a person escaping from a point on a storey actually has to travel around obstacles to reach a storey exit.

General Considerations

Effective and adequate means of escape from a shop, especially a large department store, depends, firstly, on the design and construction of the building; particularly the arrangement of spaces, the formation of compartments, the enclosure and structural protection of the escape routes and the location and installation of suitable fire doors. Secondly, when the building is in use, the contribution of management and staff; together they must ensure that fire detection, alarm and fire fighting systems are properly maintained, that goods and stock are safely displayed or properly stored and that escape routes are maintained clear of obstruction and well marked. It is also important that staff are properly trained to deal with fire emergencies including the evacuation of the chairbound disabled. Building regulations, of which the Mandatory Rules are a part, only have the power to control building design and construction. The best structural precautions may be nullified by poor management practices. Appendix A of the Code, which is not among the clauses made mandatory by the Rules, provides some advice to management about operational procedures.

The Code states that to design satisfactory means of escape requires the identification of possible sources of fire and an appreciation of the way fire develops and spreads. Of particular importance is an understanding of how smoke and hot gases, being expansive and buoyant, can travel rapidly ahead of the fire itself to inhibit the movement of those trying to escape. The inhalation of smoke or toxic gases has an immediate effect on the human brain. It can cause irrational behaviour and panic. A detailed description of the way fire develops and smoke spreads is given in clauses 4.1 and 4.2 of the Code (see also p. 2.9 of these Notes).

Provided all the recommendations of the Code are complied with, including some, for example, Appendix A, particularly A.5.4, which are not cited by the Mandatory Rules, fire should not originate in protected corridors or stairways which form the escape routes. Fire is more likely to start on selling floors, among the displayed goods, or the furnishings, or decorations; it can also start in store rooms, kitchens or offices and in rooms or ducts associated with the building services. Of these, fire starting in areas to which the customer has access will present the greatest hazard. Shops or departments which store, display and sell flammable liquids or readily ignited materials are at greatest risk.

The Code is primarily concerned with means of escape from a shop the final exit from which leads to safety in the open air. If the final exit or exits discharge into an enclosed arcade or mall, additional precautions may need to be taken to prevent fire spreading across the arcade or mall and to disperse smoke reaching the mall. The Code makes some recommendations to limit fire spread; on the subject of smoke control it recommends the BRE publication *Smoke control methods in enclosed shopping complexes of one or more storeys: a design summary* by H.P. Morgan (HMSO).

The Hazards

The main hazards to the occupants of a shop when a fire occurs are:

SMOKE

Because of the nature of the goods on display in many shops, large volumes of dense, choking, often highly toxic smoke may be released quickly in the event of fire. The sales floor of a large store generally contains a considerable amount of combustible material disposed to assist the rapid growth and spread of fire.

NUMBER OF PEOPLE

At peak shopping times and particularly during sales or special promotions, shop floor areas may be very crowded; many of the people will be unfamiliar with the layout of the shop and the location of escape routes. In an emergency people will first be reluctant to leave and then will try to leave by the route they used on entering. This can result in a milling crowd which takes longer to clear the floor area than might be estimated.

LARGE OPEN AREAS

Sales floors often cover a large area which is broken up by display stands which may lengthen considerably the distance between any point on the floor and the nearest exit. The stands also impede visibility so that the location of the nearest exit may not be obvious to a shopper on the floor.

INTERCONNECTED COMPARTMENTS

To ensure that shoppers see and circulate past as many displays of goods as possible, shop owners favour both large open areas and interconnected compartments, balconies and mezzanine floors. These features make it difficult to control the spread of smoke to confine it within compartments in the early stages of a fire.

Form of Recommendations in the Code

The Code advises on the provision of adequate means of escape from new shops. A separate section of the Code, clause 9, deals with small shops. In the event of fire, everyone within the shop should be able to escape without having to rely on outside assistance.

A small shop is defined in clause 2 as ". . . a shop with limited floor area, in one occupancy and comprising not more than three storeys". In clause 9 a maximum floor area per storey of $280m^2$ is suggested as a suitable limit; the meaning of three storeys is defined as ". . . not more than a basement, a ground and a first storey". This appears to mean that if there is no basement, a small shop should not have more than a ground and a first storey.

A shop which is to be used principally for the storage and/or sale of highly flammable liquids or materials should not be treated as a small shop. Other fire legislation may apply to it and it would be wise to consult the fire authority at an early stage of planning. There are also constraints on the location of cooking facilities in small shops and there are limits on the numbers for whom seating and standing-room may be provided.

The Code considers the escape route in two parts. Firstly, the horizontal part of the route from any point on a storey to the nearest exit from that storey, or in the case of the ground floor, the final exit. Secondly, the vertical part of the escape route comprising stairways leading to a final exit to safety outside the building.

Code recommendations are given in terms of:

1. The distance from any position which might be occupied by a shopper or a member of staff to the exit from that floor:
2. The number of routes available and their direction relative to one another.
3. The number of exits from a storey and the number of protected stairways.
4. The width of exits and stairways.
5. The siting of exits and stairways in relation to one another.
6. The capacity of inner rooms.

Travel distances, number of exits, floor area limits and exit widths recommended in the Code are aimed to achieve the orderly evacuation of an area, storey, or compartment in 2.5 minutes.

PLANNING

Site Planning

Planning multiple occupancy sites, with many shops entered off malls, arcades, etc., involves complex considerations which are not dealt with in the Code. Reference should be made to the specialised publications dealing with this subject.

When planning a shop which occupies all of a building on one site, although the Code makes no specific recommendations for site layout, complying with advice on internal planning will impose constraints on site planning. Complying with the requirements of B4 (external fire spread) of Schedule 1 of the regulations also has implications for site planning. In addition, the fire authority, using powers given them under the Fire Precautions Act 1971, may make requirements which also restrict the use which can be made of some areas of the site. To summarise, the following factors are likely to prove of particular importance.

1. The location of final exits affects site planning. Those escaping must be able to move easily away from the building to a place of safety and the route must not be impeded by, for example, security fencing or vehicles loading or unloading.

2. The permitted distance between the external walls and the site boundary depends on the materials used for constructing the walls, including the exposed outer face, and the proportion of the wall which may be "unprotected". An unprotected area is an area which comprises a window or doorway, or an area having a period of fire resistance less than that recommended as adequate (see Approved Document B2/3/4).

3. The fire authority is likely to require proper access for fire appliances and to any systems installed in the building to aid fire fighting. This will mean that certain areas of the site must be kept clear of obstruction at all times (for example, of parked cars, delivery vehicles or stock). Part of the area may have to be level hard-standing for rescue appliances.

Internal Planning

Requirement B3 of Schedule 1 of the regulations is concerned with limiting internal fire spread by structural means; B3(2) specifically imposes size limits (of building or compartment) as a method of achieving this. The Approved Document B2/3/4 suggests safe floor area and cubic capacity limits for compartments related to the period of fire resistance of enclosing structural elements. Recommendations under this heading are to be found in Approved Document B2/3/4, Table 6.1, the relevant part of which is reproduced below (Table 3.1).

Table 3.1 Shops–size and fire resistance (Extracted from Approved Document B2/3/4, Section 6, Table 6.1)

TYPE OF BUILDING	SIZE LIMITATIONS			FIRE RESISTANCE (Hours)	
	Ht (m)	Max. floor area (m^2)	Capacity (m^3)	Above ground	Basement and floor over
Single storey	–	2000	unlimited	½	n.a.
	–	3000	unlimited	1	n.a.
	–	unlimited	unlimited	2	n.a.
Multi-storey no sprinklers	7.5	500	unlimited	½	1
	15	unlimited	3500	1	1
	28	1000	7000	1	2
	n.l.	2000	7000	2	4
Multi-storey sprinklered	7.5	500	unlimited	½	1
	15	unlimited	3500	1	1
	28	1000	7000	1	2
	n.l.	4000	14000	2	4

A second factor affecting internal planning and the size of compartments is the maximum distance an occupant can be expected to travel across an open floor area to reach the nearest exit or exits giving on to a protected escape route. More information about maximum travel distances will be found in paragraph *(b) Distance* below and in Table 3.2

The Code describes three quite different internal planning arrangements and makes different recommendations for means of escape for each. The different arrangements are:

1. Compartmented: A shop may occupy several storeys but be constructed so that each storey is a separate compartment. Openings through the compartment floors for accommodation stairways, escalators, lifts, etc., which allow shoppers to circulate freely between storeys, are permitted provided they are fitted with steel fire doors or shutters held open by fusible links. Since these will not close soon enough to arrest the spread of smoke between compartments, other features are recommended (for example, fire screens and self-closing fire doors) to control smoke movement. Otherwise all interconnected storeys lacking these features must, for means of escape purposes, be treated as one compartment.

2. Open spatial (vertical) planning: It is considered good for marketing to arrange for customers to be able to overlook lower sales floors from higher storeys so that they can see displayed goods. This may be achieved by planning storeys around a courtyard or open central well. Alternatively, a split-level floor plan may be adopted. When planning means of escape, shops with such open plan arrangements may have to be treated as one large compartment.

3. Open storey (horizontal) planning: In this case the whole sales floor area of a storey is open and has no partitions or screens to interrupt sight-lines across the storey. With such an arrangement, although the entire floor may be large, shoppers and staff would be quickly aware of an outbreak of fire anywhere on that storey and know in what direction to go to reach a convenient exit. At the same time it must be appreciated that the open plan permits more rapid spread of fire and smoke.

RECOMMENDATIONS FOR SHOPS OTHER THAN SMALL SHOPS

The Horizontal Part of the Escape Route

In designing for this part of the escape route, account must be taken of–

(a) the maximum number of people likely to be on each storey,
(b) the distance each person should have to travel to reach an exit,
(c) the need to provide alternative routes to separate exits,
(d) the width of each exit and the aggregate width of all exits,
(e) the minimum number of exits from a storey.

(a) Number of People: In the case of small rooms or rooms with seating accommodation, the number of occupants will be known and exit width designed for that number. Elsewhere, numbers are calculated using an occupancy factor which is related to the use of the area in question. Table 3.4 gives the factors recommended by the Code; also shown in the Table are the maximum numbers of people who can occupy typical areas of floor. The areas chosen for this Table correspond to the area limits of Table 3.1

(b) Distance: The Code recommends the use of two distance limits when planning the layout of a storey, particularly the location of exits. The first is the distance a person would actually have to travel from any point on the floor to the nearest exit, moving around all obstacles such as counters and display stands and through openings in partitions. The Code calls this "the travel distance" and recommends a maximum value.

Because, when the building is being designed, the positions of shop fittings and similar obstacles are unlikely to be known, the Code makes use of a second measure. This is the straight-line distance from a point on a sales floor to an exit ignoring all obstacles; this is called "the direct distance". Figure 3.1 illustrates the two "distance" concepts.

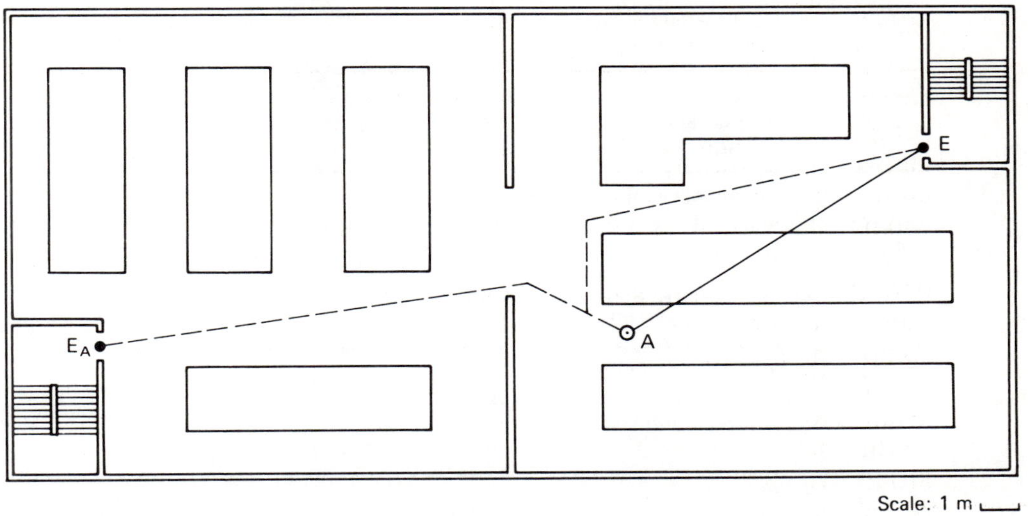

From point A -
AE (full line) = DIRECT DISTANCE to storey exit
AE (broken line) = TRAVEL DISTANCE to storey exit
AE_A (broken line) = alternative route to a second exit

Notes

- Measurements are taken to the centre of the exit.
- Line of travel allows adequate clearance around obstacles.
- Maximum travel distance is 1.5 times maximum direct distance.
- At the scale used, direct distance AE is about 9.5 m, well under the permitted maximum of 30 m (see Table 3.2).
- Total floor area (excluding protected stairways) is about 306 m^2; with a floor factor of 7.0 for a general shop (Table 3.3) the number of people on this storey is 43. This number can pass through one exit of minimum width 826 mm (Fig. 3.4) and one stairway width of 1000 mm (Fig. 3.5). The second stairway permits one to be discounted.

FIG.3.1 Direct distance and travel distance: possible arrangement of sales floor with counter/display areas

The means of escape requirement of B1 applies, in the case of a shop, to a new building; thus the recommendations of the Code have to be complied with at the design stage when the positions of fittings, etc. (and therefore travel distances) are unlikely to be known. In these notes, therefore, only direct distance limits are given in the Tables and used in the drawings which illustrate Code recommendations. It must be remembered, however, that "travel distance", will have to be checked before the building can be occupied; "travel distance" is 1.5 × "direct distance".

The Code recommends two values for direct distance; the lower figure (shorter distance) applies when travel is possible in only one direction–in an inner room or to a single exit in a "dead end" situation. The Code also gives a higher figure which may be used when two or more exits are available, and the routes to them are in directions which diverge enough to be considered to offer "alternative means of escape" (see below).

The Code recommendations for maximum direct distance from various situations on a storey to an exit are given in Table 3.2.

Distance limitations, and the need to locate exits strategically around the perimeter of a floor to achieve the recommended angular separation of routes, places constraints on the planning of a storey and the location of exits. The location of exits in turn dictates the placing of stairways.

Table 3.2 Maximum recommended direct distances in a storey

LARGE SHOP–One in which the floor area on any storey exceeds 280m^2	
Situation and escape option	Maximum direct distance (m) to the only or nearest exit
Any storey* from which escape is possible in only one direction–"dead-end"	12
Any storey* having more than one exit but routes less than 45° apart and without fire resisting separation between the routes	12
Any storey having more than one exit, with all routes less than 45° apart separated by fire resisting construction	30 which may include up to 12m of "dead-end"
Any storey having more than one exit and with all routes diverging at 45° or more	30 which may include up to 12m of "dead-end"
Ground storey with open plan sales area with exits not more than 60m apart evenly distributed around it and visible from all parts	45 which may include up to 18m of "dead-end"

* Not in a building with open spatial (vertical) planning.

Table 3.3 Maximum recommended direct distances in a storey

SMALL SHOP–One with a floor area on a storey not greater than 280m^2	
Situation and escape option	Maximum direct distance (m) to the only or nearest exit
Ground storey only one exit	18
Basement or first storey exit by single stairway	12
Any storey with more than one exit or stairway	30

(c) Alternative routes: It is the general rule that alternative routes, offering escape in significantly different directions, should be available from any point on a storey. This enables a person faced with an outbreak of fire to escape by moving away from it. "Significantly different directions" means in the Code that the lines directly connecting the point to two exits should enclose an angle of 45° or more unless the routes are separated for at least 60 per cent of their length by fire resisting construction (see Figs 3.2 and 3.3).

(d) Width: Two "widths" are important: the width of an exit or escape route, and the aggregate width of all exits, etc., which serve a storey. The width recommendations in the Code are based on experimental data gathered many years ago; several factors influence rate of movement of people through an exit, but an average of 40 persons per minute through a unit width of 500mm is the standard generally adopted. Fortunately, to increase capacity, exit widths do not have to be increased in 500mm steps. Observation shows that small increases in width allow files of moving people to overlap and so increase exit capacity.

The relationship between exit width and the number of people able to pass through it in the standard escape time is shown graphically in Fig. 3.4. The Code provides this information in a table with widths increasing in 100mm steps; it is accepted that intermediate values can be used to suit standard door widths.

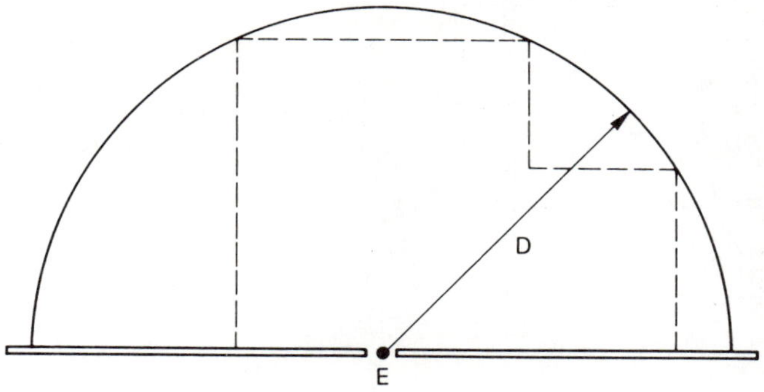

In this example there is a single exit E; no part of the floor area may lie outside the semi-circle drawn with its centre at the exit and radius D. Distance D is the MAXIMUM PERMITTED DIRECT DISTANCE. The broken line shows one acceptable floor area.

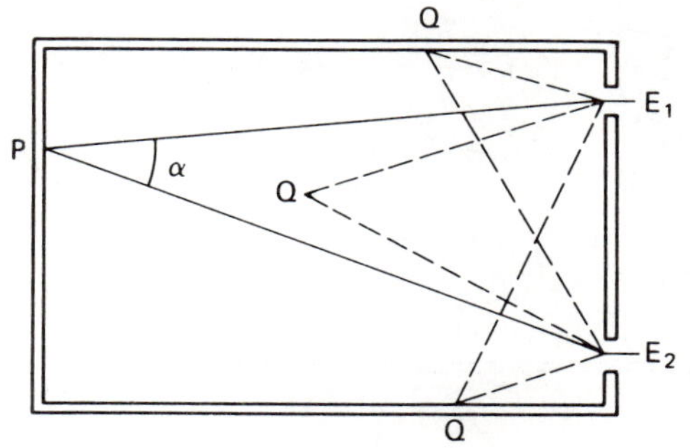

In this example there are two exits E_1 and E_2 but from any point P within a large area of the floor the angle α contained by the direct lines to E_1 and E_2 is less than 45°. (The angles E_1QE_2 define the 45° limit.)

In this example, therefore, the MAXIMUM PERMITTED DIRECT DISTANCE from either exit is the same as for an area or storey with a single exit.

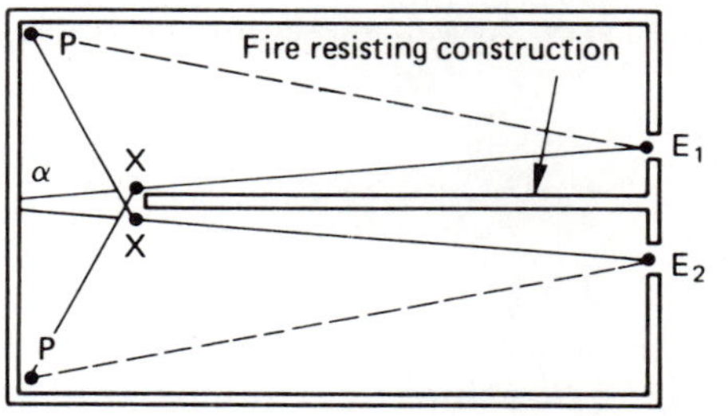

In this example there are two exits E_1 and E_2; a fire resisting wall partially divides the floor area. Although angle α is less than 45°, the routes to the two exits are separated by the wall for more than 60% of the distance From points 'P', the distance PX may not exceed 12 m: the total distance PXE should not be greater than 30 m.

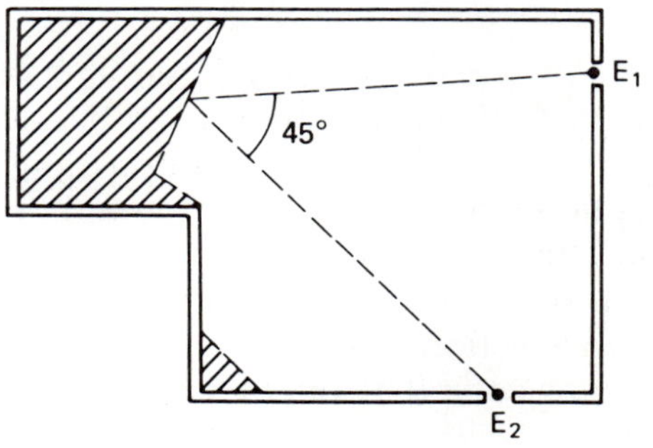

In this example there are two exits E_1 and E_2. From any point on the unshaded part of the floor, the lines of direct distance diverge by 45° or more. Travel through the shaded areas by the shortest route to the edge of the unshaded area must not exceed 12 m and the total distance from any point to an exit shall be not more than 30 m.

FIG.3.2 Direct distance and direction

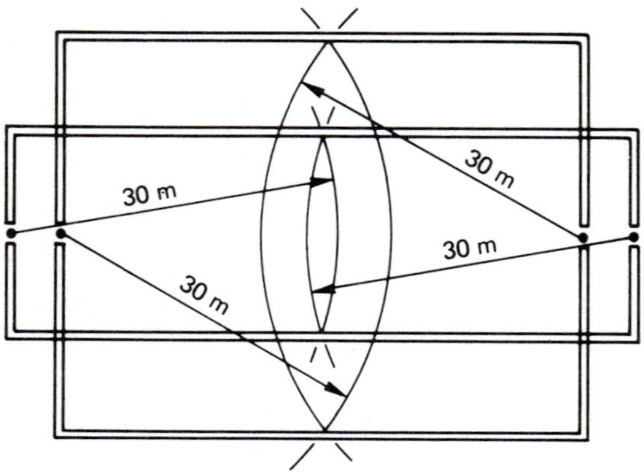

Two floor plans with exits at the mid-point of each end and with all points of the floor not more than 30 m from an exit.

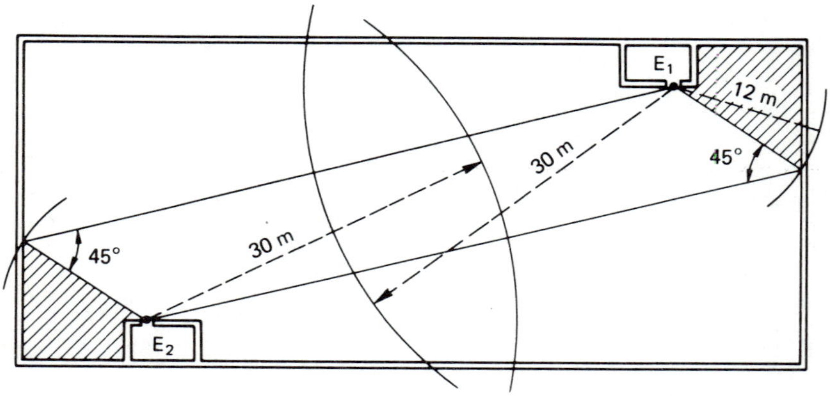

In this example there are two exits E_1 and E_2. From any point in the unshaded area, lines representing direct distances to E_1 and E_2 enclose an angle of 45° or more. The limiting points from which the lines diverge at exactly 45° are shown.

Thus, from within the shaded area, the MAXIMUM PERMITTED DIRECT DISTANCE may not exceed 12 m. Elsewhere every point should be not more than 30 m from one of the exits. The 30 m limits are shown to overlap comfortably.

FIG.3.3 Alternative means of escape

(e) *Minimum number of exits:* Only small rooms and restricted "dead-end" areas may be served by one exit. Elsewhere, the minimum number will be dictated by travel distance limitations and the need to provide alternative routes. In addition, the number must allow for any exit being rendered unusable by fire in its vicinity.

(f) *Escape by way of a flat roof:* In the design of a new building it should not be necessary to use a flat roof as part of an escape route. There may be circumstances, for example, on a very confined site, when it is impossible to avoid; but these should be regarded as exceptional and every effort should be made to contain the escape route within the building.

Escape from an area via a flat roof is only acceptable when there is an alternative route from the same area. In addition–

- the roof must be part of the building served by the escape route,
- the route should lead to another storey exit,
- the route should be adequately marked and guarded in accordance with BS 6180: 1982,
- the part of the roof forming the escape route and the structure supporting it should be constructed as a fire-resisting floor, and
- the route should be well away from ventilators and non fire-resisting rooflights, windows and doors and any other opening which could make it unusable were smoke or fire to issue from it.

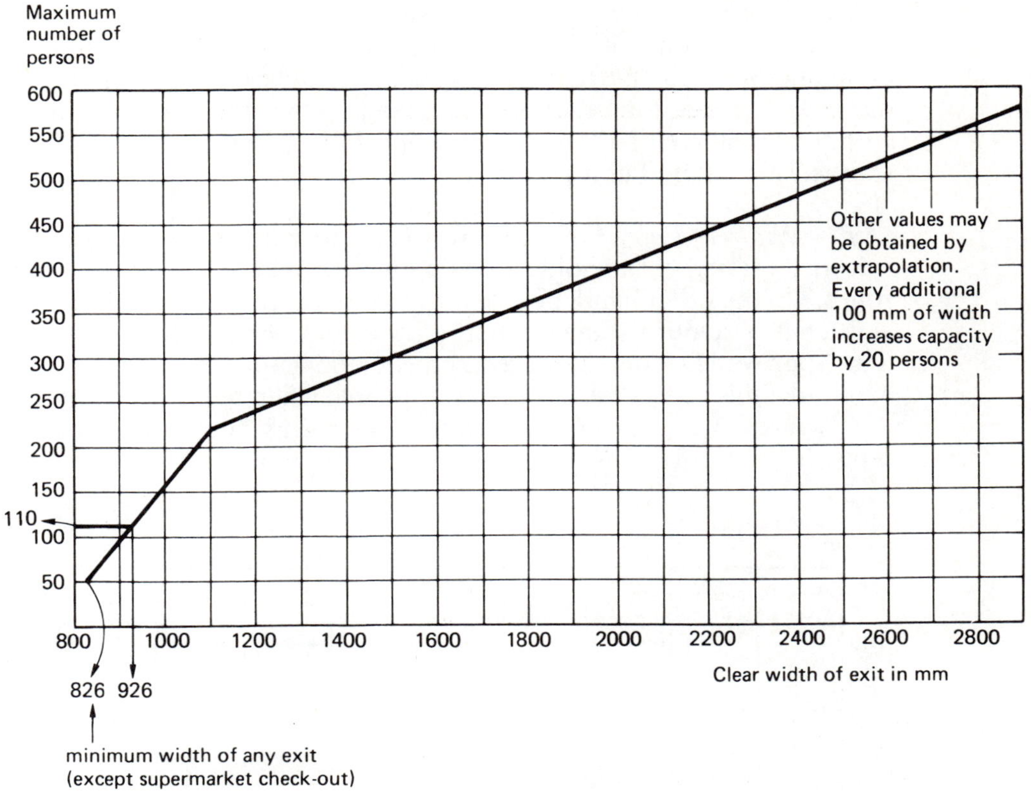

FIG.3.4 Capacity of storey exits (persons per 2.5 minutes)

(g) Inner rooms: Inner rooms should not be designed to accommodate more than 50 people. Escape will be through an access room (see Fig. 3.9). The travel distance limit appropriate to a "dead-end" applies and is measured from a point most remote from the door to the access room, across to the access room exit by the shortest route and onwards to a doorway leading to a protected zone. The access room should not be a room with a high fire risk. Except in the case of a fitting room, a person in the inner room should have a clear view of the access room through glazing in the door or wall (or both).

The Vertical Part of the Escape Route–Stairways

Lifts, escalators and open accommodation stairs are discounted as means of escape (except that a lift or lifts may be designated for the escape of chairborne disabled persons). Only protected stairways can be used as the vertical part of the escape route. Such stairways are enclosed in fire resisting construction and are separated from the horizontal part of each storey by self-closing fire doors and in some circumstances by protected approach lobbies. Paragraphs 6.29 to 6.34 and 6.42 to 6.44 of Approved Document B2/3/4 give detailed recommendations. There are restrictions on what the enclosures may contain by way of services, cupboards, etc. and on the location of windows.

(a) Siting of stairways: The siting of stairway enclosures must take account of the exit requirements on all storeys served and the need to afford effective alternative directions of escape (see above). It is also necessary to ensure that the final exit at the foot of the stairway discharges to a place of safety (see site planning). It is important that storey exits leading to stairways are sited as far away as possible from the heads of escalators or accommodation stairways.

(b) Number of stairways: The numbers of stairways required will depend on whether the building is compartmented at every floor or has some vertical open planning. With compartmenting, stairway capacity is based on the simultaneous discharge of two storeys; with open planning the stairways must be able to cope with the discharge of all storeys together. In all cases it must be assumed that one stairway is rendered unusable by the fire, thus the minimum number is two.

(c) Width of stairways: Stairway capacity is related both to width and to the fact that a stairway can hold a certain number of people in each of its inter-storey zones. On the other hand, people move more cautiously on stairs than on the level and tend to favour the sides where they can use handrailing. It is for this reason that Approved Document K1 recommends 1800mm as a maximum width between handrails; this allows for three files of people when being used at full capacity.

Figures 3.5 and 3.6 show the relationship between stairway width and capacity.

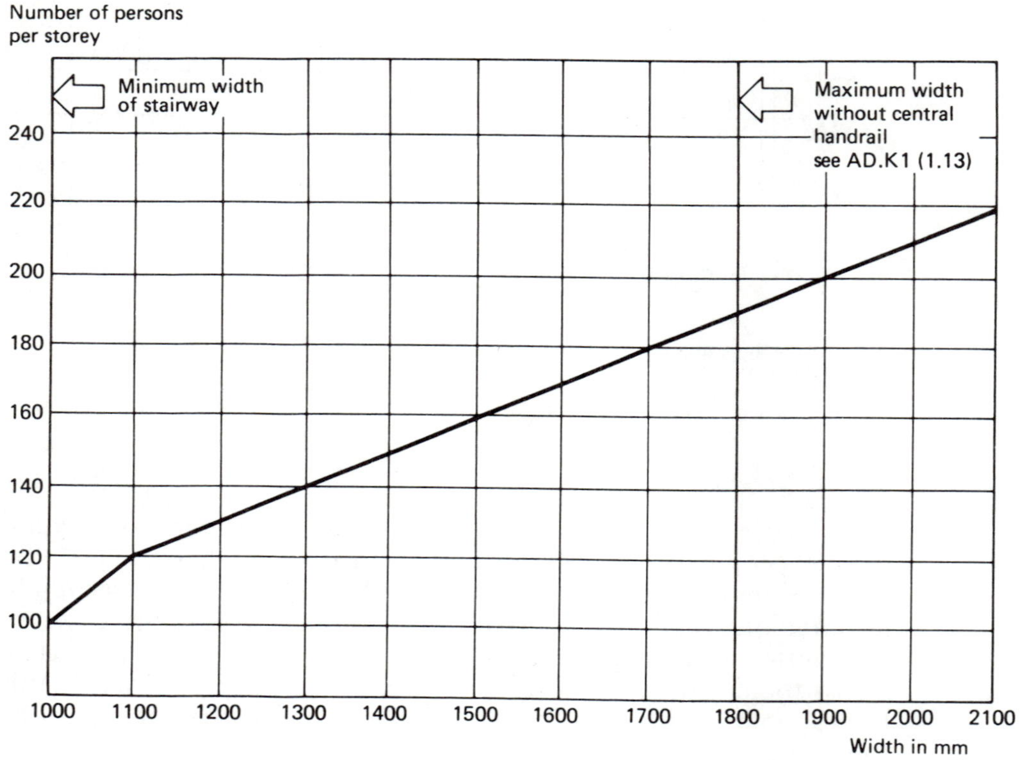

FIG.3.5 Stairway width for the evacuation of two adjacent storeys in a horizontally compartmented building

(d) Enclosure of stairways: The fire resistance of the enclosure should be not less than 30 minutes; with large shops Approved Document B2/3/4 will recommend longer periods. The enclosure may contain sanitary accommodation provided it is not used as a cloakroom and does not contain a portable heating appliance or any gas appliance other than a water heater or incinerator. Cupboards may be located within the enclosure provided they are enclosed in fire-resisting construction. Every protected stairway should have its own enclosure and should have only one entry point on each storey.

If a protected stairway is recessed behind the line of the elevation or projects in front of it, it is important to ensure that fire in the building cannot make the stairway unusable because of the juxtaposition of windows or any other opening. A minimum distance of 1.8m should be maintained between windows at the same level in walls at rightangles to one another. If a window in the building is within 9m vertically below a stairway window and within 1.8m horizontally, the former should have fixed fire-resisting glazing.

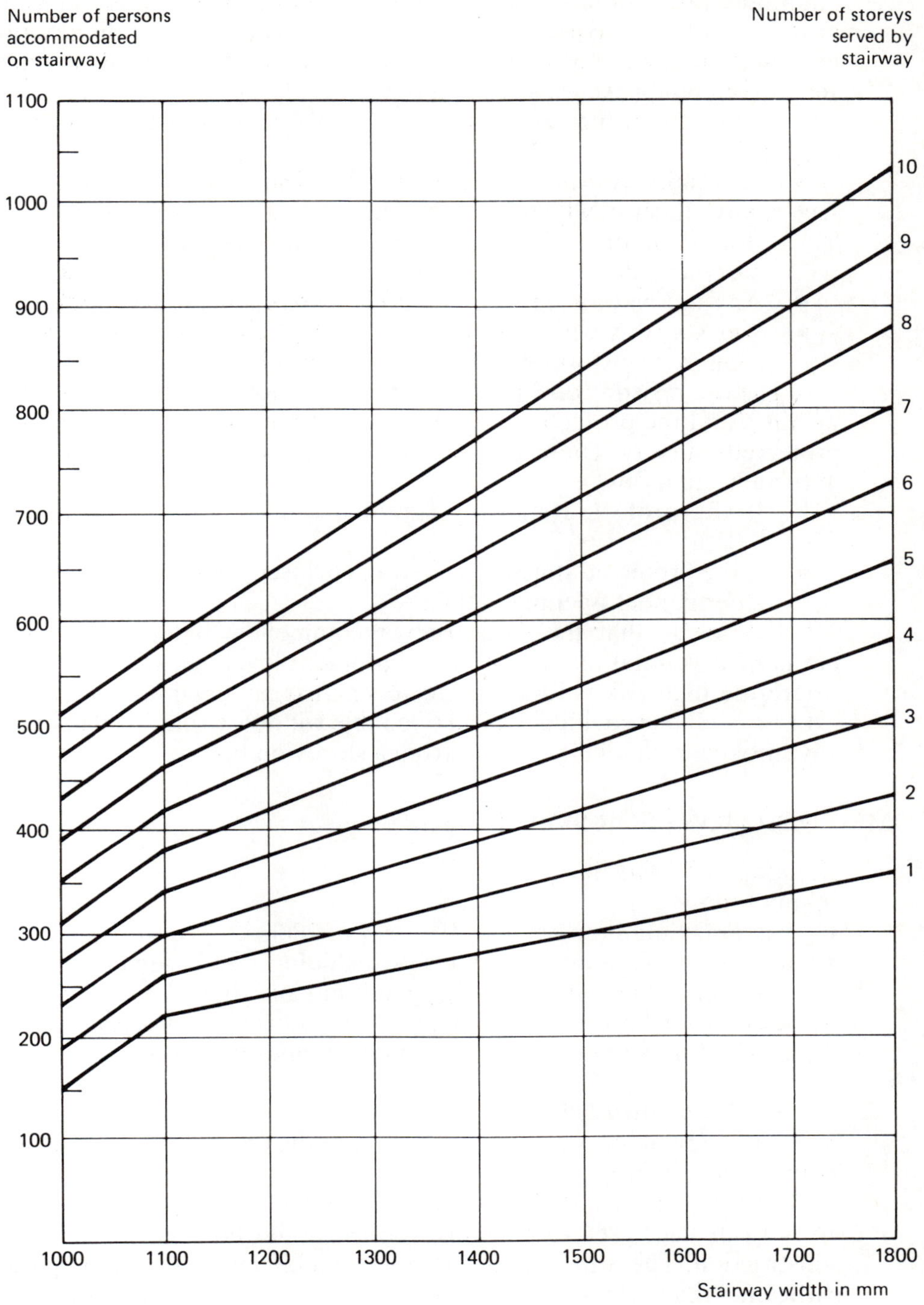

FIG.3.6 Stairway width for the evacuation of basements and for the total evacuation of the building

(e) Stairways and basement storeys: Ideally, stairways giving access to basements should be entered at ground level from outside the building. It is permitted, however, in buildings which have two or more stairways for one or more to continue down to the basement. But there should always be one stairway serving the upper floors which finishes at ground level and does not extend to the basement. All stairways which do connect upper storeys with a basement should have a protected lobby between the stairway and every basement level.

(f) Protected lobbies: In some circumstances, stairway enclosures need additional protection against the entry of smoke. This is done by constructing a protected lobby or corridor between every storey exit and the doorway leading into the stairway enclosure on all of the upper storeys. The only exception is a top storey containing exclusively plant rooms.

In the following circumstances the Code recommends protected lobby approach:

- Any stairway in a building greater than 18m in height, at every upper storey.
- Any stairway which connects the ground or upper storeys to a basement storey, or one which serves more than one basement storey, at every basement storey.
- Any stairway giving access to an enclosed car park, at every car park level.
- Any stairway which directly serves a high fire risk area, for example boiler room or transformer chamber, at the level of that risk.

The fire resisting enclosure for a protected lobby, the services, etc. permitted within it and the protection of the opening giving access to it is the same as for a protected stairway. There are also recommendations for the lobby to be ventilated in certain situations:

- If the protected stairway serves a basement, a permanent opening of $0.05m^2$.
- If the protected stairway serves an enclosed car park or an area of high risk, a permanent opening of $0.4m^2$.

It appears that the Code only recommends that provision be made for permanent ventilation of lobbies at basement level, car park level and the level where the high risk is located. However, it is an advantage if all lobbies on an external wall be provided with an openable window to allow them to be cleared of any smoke which might creep in from a storey on fire.

RECOMMENDATIONS FOR SMALL SHOPS

Definition of Small Shop

A small shop–

- has a maximum floor area per storey of $280m^2$.
- has not more than three storeys including any basement; if there is no basement, there should only be a ground and first storey.
- should not be used principally for the storage and/or sale of hazardous goods (for example, highly flammable liquids or solids).

General Recommendations

The sales area of any storey should be generally undivided so that exits are clearly visible from all parts.

Offices, stores and kitchens may be divided off and the latter should be sited away from sales area exits. Store rooms should be enclosed with fire resisting construction. The wall dividing an office or a kitchen from the shop should have a sufficient area of clear glazing to give an occupant immediate warning of fire in the shop.

If the small shop is (or contains) a bar or restaurant, seating accommodation or assessed standing accommodation should not exceed 30 persons per storey (see Table 3.4 for floor space factors). If the bar or restaurant is on the ground storey and that storey has its own independent final exit(s), seating or standing for 100 is allowed.

Direct Distance to Storey Exit and Number of Exits

Maximum direct distances applicable in small shops are set out in Table 3.3. When a storey has more than one exit so that alternative routes are available, the same 45° separation between the routes is required as is the case in large shops (see Figs. 3.2 and 3.3). In small shops with open stairways, distances in the basement are measured to the foot, and on the first floor to the head of the nearest stairway.

TABLE 3.4 Floor space factors and numbers of occupants

Description of room or storey	Floor area per person m^2	Maximum number of people in typical areas				
		Small Shops		Others		
		170m²	500m²	1000m²	2000m²	4000m²
SALES AREAS						
Main sales areas of department stores, supermarkets, covered markets and bazaar type shops	2.0			500	1000	2000
General shops and showrooms and those areas of department stores likely to be sparsely occupied.	7.0	24	71	142	285	571
CATERING FACILITES						
Bar (Drink or snack mainly consumed standing)	0.3	566				
Restaurant or lounge	1.0	170		1000		
Kitchen, etc., areas	7.0					
STAFF ROOMS	1.0					
STORAGE						
Goods, car parking*	30.0					

* Alternatively, 2 persons per parking space

SMALL SHOP: small, two-storey, with single protected stairway (may comprise a bar or restaurant).

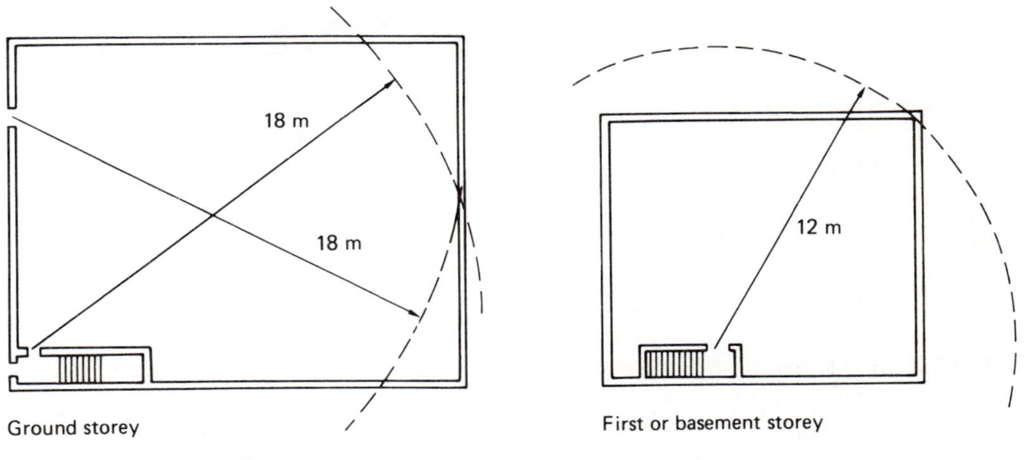

Notes

1. Fire doors should open in the direction of escape unless they serve an area which will never accommodate more than 50 people.
2. The ground storey exit arrangement and the limit of 18 m direct distance limits the floor area to some 200-250 m^2. At 7.0 m^2 per person floor space factor, this allows a maximum of 28-36 persons.
3. This diagram is based on Fig. 5 of the Code but note that the Code figure shows the ground storey plan drawn to a different scale from the plan of the first storey (or basement). This means that the floor area of the ground storey is greater than either of the others.
4. Note also that Fig. 5 of the Code '. . . two or three-storey shop'; the stairway arrangement shown cannot, however, serve both a basement and a first storey.

FIG.3.7 Horizontal travel – direct distances

Stairways

It is always preferable for stairways in small shops to be protected and to discharge directly to a final exit. This applies without exception to all stairways in a small shop which comprises a bar or restaurant.

In other cases, when a storey is served by one protected stairway, a second stairway may be open (unprotected).

In a very small shop (one in which none of the storeys exceeds $90m^2$ in area) which has a single stairway connecting the upper floor, or the basement, or both to the ground floor, one stairway may be open provided–

- the open stairway delivers into the ground storey at a point not more than 3m from a final exit; and
- (if there are three storeys) the other stairway is protected and discharges to its own final exit.

These rules and the direct distances recommended by the Code are illustrated in Figs. 3.7 and 3.8.

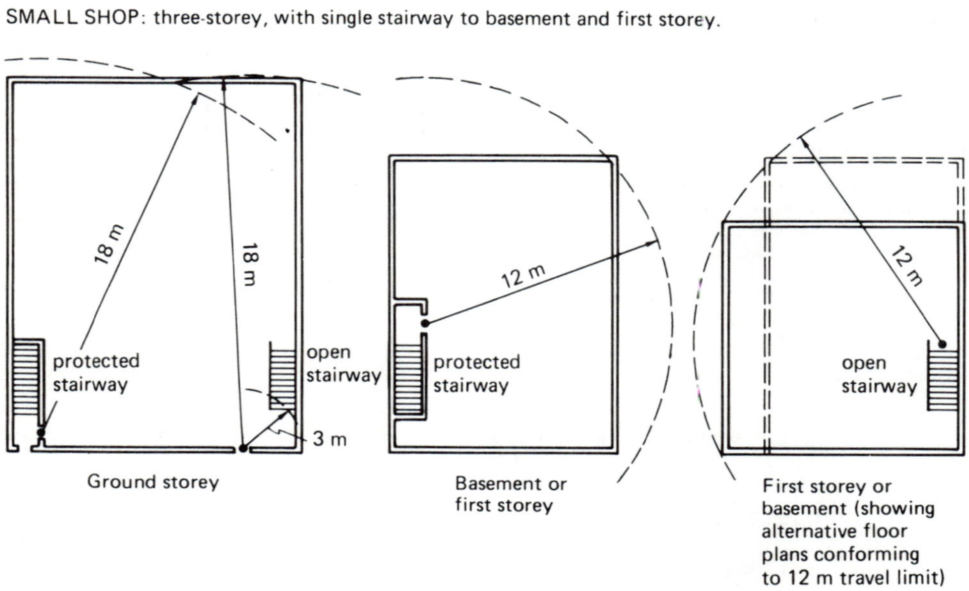

FIG.3.8 Direct distances

CONSTRUCTION

General

Buildings which conform to the requirements B2, B3 and B4 of the Schedule to the building regulations will easily satisfy the recommendations of the Code for fire resistance, etc. However, the Code emphasises two additional points. The first concerns the selection of building methods and materials for fire-resisting elements which can withstand mechanical or other damage in use, so that the fire-resisting quality is not destroyed or reduced. The second point concerns junctions and joints; care should be taken to avoid weaknesses at junctions between elements (for example, wall to ceiling), and at joints between sheet materials which provide fire protection to an element of structure. The Code also recommends care in choosing the method of fixing sheet materials so that they remain in position when subjected to fire.

Glazed Elements

Areas of glazing in walls intended to protect an escape route or enclose a stairway can present a hazard to people escaping if they do not prevent heat radiation. Some glazing, when subjected to the standard fire test, can provide resistance in terms of both integrity and insulation for the half-hour period which is sufficient for means of escape. Most glass and glazing, however, cannot meet the insulation requirement for more than a few minutes and the Code recommends the following restrictions in its use:

In any wall separating a protected zone from an unprotected area, such glazing should only be used at a height of 1.1m above finished floor level. This allows people escaping to crawl past if heat is excessive.

In a fire-resisting door between a protected stairway and the sales floor or unprotected corridor, only 50 per cent of the area of the door may be glazed.

Fire Doors

Fire doors are classified by reference to the period in minutes during which their integrity is maintained under test. Thus a door is designated, for example, FD 20, if integrity is sustained for 20 minutes. Where doors are, in addition, required to retard the passage of smoke at ambient temperature, the suffix "S" is added; this generally means that it will be fitted with a flexible edge seal.

Location of door	Recommended designation
In a protected stairway enclosure	FD 30S
In a protected lobby or corridor enclosure	FD 30S
In an opening serving normal hazard areas (store, kitchen, etc.)	FD 30
In the enclosure of a high hazard area	FD 60
In lift shafts opening into unprotected areas	FD 30
In openings into ducts, etc.	FD 30
Across corridors more than 12m long which are part of an escape route	FD 20S
Protecting a dead-end corridor	FD 20S

All the above fire doors, except those to cupboards or service ducts, should be self-closing; rising butt hinges cannot be relied on to provide this function. The self-closing device should not be one which can easily be disconnected or immobilised and it should be powerful enough to override any latches; it should be impossible to retain the doors in the open position.

Where necessary, an electromagnetic or similar device may be provided to hold a door open against the self-closing device but the door must be released automatically by the smoke detector system, a manual alarm system, by manual operation at a central control or by power failure.

Cupboard doors or service duct covers which are not required to be self-closing should be kept locked shut when not being used; they should be marked on the outside "KEEP LOCKED".

Hinges for fire doors should be non combustible and should not have a melting point below 800°C. Other door furniture should be simple and easily operated from the escape side; Panic bolts should comply with BS 5725: Part 1; any other security device should have fire authority approval.

Construction of Escape Routes

Important features are:

(a) The use of non-slip floor finishes.
(b) Adequate headroom–not less than 2m.
(c) Freedom from wall projections which could catch on clothing or otherwise impede the flow of people using the route.

(d) Ramp gradients no steeper than 1 in 12.
(e) Stairways constructed in accordance with Part K of Schedule 1 of the regulations in respect of going, handrailing, balustrades and headroom. The treads should have a non-slip finish.

ENGINEERING SERVICES AND ANCILLARY ACCOMMODATION

Introduction

Engineering services and ancillary accommodation are important from the point of view of means of escape for two main reasons. First, the potential fire hazard some present to building occupants and the escape routes provided for their safety. Second, the threat they may present to staff working in the accommodation or directly involved with the service installations.

The Code lists all engineering services but only some have implications for means of escape. They are:

(a) Protected electrical circuits which serve emergency lighting, alarm systems or other fire related equipment.
(b) Emergency lighting of escape routes.
(c) Boiler rooms and fuel storage.
(d) Ventilation and air conditioning systems.
(e) Protected stairway pressurisation systems.
(f) Lift motor rooms.
(g) Refuse disposal systems.
(h) Refrigeration rooms.

Ancillary accommodation is considered under two heads.

On the one hand there are those rooms or areas which present no greater hazard than normal sales floor areas. These are usually occupied by sales staff and possibly the public and include staff rest rooms, office accommodation, fitting rooms, and telephone exchanges. Large office areas should satisfy the requirements of Section 4 of the Mandatory Rules.

Considered separately are what the Code describe as "higher fire risk areas"; these are defined as "areas with a fire risk higher than that of a sales area". The Code lists in this category rooms associated with engineering services like boiler rooms and transformer chambers, storage areas, loading bays, refuse rooms, car parks, workshops and kitchens.

If the construction of rooms, chambers and other parts of the building associated with building services is carried out in accordance with the requirements of Parts B2 and B3 of Schedule 1 of the regulations the structural recommendations of the Code will have been met. In addition the Code recommends that the installation of services should be carried out to conform with the appropriate British Standard. Those quoted are as follows:

Lift motor rooms	BS 5655: Pt 1: 1979
Waste storage chambers, chutes and hoppers	BS 5906: 1980
Pressurisation systems	BS 5588: Pt 4: 1978
Mechanical ventilation and air conditioning	BS 5720: 1979
Service and ventilation ducts	BS CP 413: 1973
Oil-fired installations	BS 5410: Pt 1: 1977
	: Pt 2: 1978
Oil storage tanks	BS 779: Pt 5: 1975
Town and natural gas installations	BS 5376: Pt 2: 1976
	BS CP 332: Pt 3: 1970
Escape route lighting	BS 5266: Pt 1: 1975

Ancillary Accommodation–Means of Escape

The means of escape from normal fire risk areas should follow the recommendations for sales floor areas using "travel distance" rather than "direct distance". Within rooms, travel will generally be possible in one direction only (that is, the room will have only one door). None of the ancillary accommodation should open directly into a protected stairway enclosure.

In higher fire risk areas the portion of the travel distance which is within the room is further restricted. If escape is possible in only one direction within the room, there should be not more than 6m to the room exit. If escape is possible in more than one direction, 12m is permitted.

Inner Rooms

An inner room should not be planned to accommodate more than 50 people; the escape route from it should not have to pass through more than one access room which should not be "a room of higher fire risk". There should be clear glazing in some part of the enclosure to allow occupants a view of the access room.

Corridors

Corridors serving ancillary accommodations should be enclosed by construction with half-hour fire resistance; all doors should be fire resisting and self-closing. Corridors longer than 12m should be fitted with doors to subdivide them and restrict the spread of smoke. The distance between doors will be determined by travel restrictions as shown in Fig. 3.9.

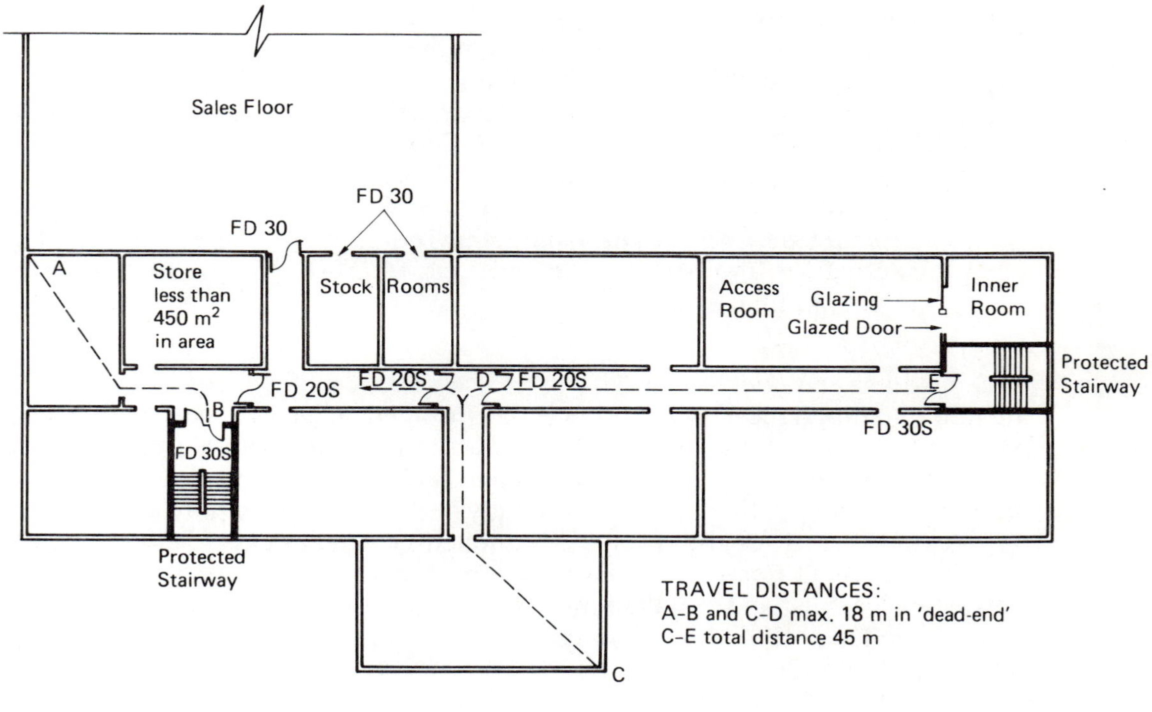

FIG.3.9 Ancillary accommodation: division of corridors, fire doors, etc.

Service Accommodation of Higher Fire Risk

Any room in which flammable liquids or gases are stored or used should have a cill at the doorway and an interceptor for any floor drainage. All service accommodation should have adequate ventilation to the external air. The exit from service accommodation situated at basement level should not be sited so as to prejudice escape through any final exit from the building.

Higher fire risk areas should have a protected lobby or corridor between them and any protected stairway forming part of an escape route.

Walk-in Refrigerated Cold Rooms

The Code gives quite detailed recommendations for the construction of cold rooms, particularly with regard to insulation materials. There is a Standard dealing with this subject (not referred to in the Code); it is BS 2502: 1979 "Specification for manufacture of sectional cold rooms (walk-in type)". The Code recommends consulting the fire authority concerning suitable fire access points.

Transformer and Switchgear Rooms

Unless situated on the roof or away from the main building, these rooms are best sited on an external wall and entered from outside the building. There should be adequate ventilation and the travel distance within the room limited to 6m if there is one exit, otherwise 12m.

Protected Circuits

These should be provided for all equipment which contributes to safety in the event of fire, notably alarms, detectors and escape route lighting. The circuits should consist of mineral insulated metal sheathed or other heat resisting cable. The route chosen for the wiring should not pass through areas with a significant fire risk and the circuits should be independent of any other wiring.

Escape Route Lighting

The designer should ensure that all occupied accommodation, corridors and protected stairways will be adequately lit day and night. This applies particularly to underground or windowless rooms and central core stairways. Some internal corridors may be without natural light.

The Code excepts from this requirement the following:

(a) Corridors which do not exceed 30m in length.
(b) Stairways in buildings less than 18m in height particularly where there is adequate street lighting and no sales floor exceeds $280m^2$ in area.
(c) Ground storey sales areas not greater than $280m^2$ in area provided there is nowhere more than 15m from the nearest final exit.
(d) Ground storey restaurant, banqueting room, public house, bar or similar catering for fewer than 100 people and from which there is a sufficient number of final exits.

Boiler Rooms and Fuel Storage

In new buildings these will normally be installed in accordance with British Standards. Any boiler rooms which are not should have adequate provision for smoke venting and in both boiler rooms and fuel storage rooms, travel distances should not be longer than 6m if there is only one exit or 12m with two or more.

Arrangements for the storage of liquefied petroleum gas should follow the guidance in Health and Safety Executive note CS 5, "The storage of liquefied petroleum gas at fixed installations". If the gas is for an internal combustion engine, the latter should be surrounded by a bund wall not less than 600mm in height above floor level.

Ventilation Systems

Ventilation systems should be designed so that the normal airflow pattern is away from protected escape routes and exits. If the system recirculates air, it should be provided with smoke detectors in the extract ducts which, when smoke is present, will direct extract air to the outside of the building and stop recirculation. The system should also be compatible with any pressurisation system fitted.

Lifts

Lifts are not to be used as part of the designed means of escape except that one or more may be designated for use in emergency by the chairborne disabled. Fire-fighting lifts are outside the scope of the Mandatory Rules.

Lift motor rooms are better located above the lift well; if it is necessary for them to be at the foot of the lift shaft, fire separation between the motor room and the shaft is necessary. The lift shaft will have to be constructed as a protected shaft to comply with the requirement B3 of Schedule 1. Lift motor rooms should comply with BS 5655.

Waste Disposal
A refuse chute is within the scope of the definition of protected shaft given in Appendix L of Approved Document B2/3/4; recommendations for the construction of protected shafts will be found in paras. 6.29 to 6.34 of the Approved Document.

Requirement H4 of Schedule 1 of the regulations requires the provision of satisfactory means of storing solid waste in a building and for there to be adequate access from the street for its removal. Recommendations for satisfying these requirements will be found in Approved Document H4.

The Approved Document and the Code for shops both cite BS 5906: 1980, which is the Code of Practice for storage and on-site treatment of solid waste from buildings.

Incinerators
Two types of incinerator may be installed in shop premises; they may be sanitary incinerators in toilet accommodation or an incinerator for disposing of combustible bulk waste. Either type may be fired by gas or electricity.

As both types of incinerator are "heat producing appliances" the installation and associated flues are subject to requirements J2 and J3 of Schedule 1 of the regulations. General recommendations on how to comply are to be found in Approved Document J1/2/3 although this document does not specifically mention incinerators. The only recommendation of significance in the Code is that if capacity exceeds $0.08m^2$ it is preferable that the incinerator be located in a separate building.

Storage, etc., of Goods
Rooms where goods are received, sorted, stored and prepared for dispatch present a fire hazard; loading bays come into the same category. An area involving storage of goods is better situated on an external wall and in a location where smoke discharging from it cannot jeopardise an escape route from the building. There should be adequate fire resistance in the structure separating these areas from the remainder of the building; the recommended period of fire resistance will be higher when the goods stored or handled are of a hazardous nature.

The Code also recommends that any main storage area which
(a) is situated below ground, or
(b) exceeds $450m^2$ in area
should have adequate provision for smoke venting.

Car Parks
There are requirements under the Petroleum Act controlling various aspects of the design of large car parks and underground car parks. The Code recommendations which relate to means of escape are as follows:

If any car park is within or adjoins a shop and there is direct access between the two, the access should incorporate a protected lobby.

An opening provided for lighting or for ventilating the car park should be so situated that it does not endanger any escape route or the final exit from any part of the shop.

There should be adequate provision for venting smoke.

Workshops
The Code makes no recommendations beyond noting that if flammable materials are involved, special requirements may be imposed by the fire authority or inspectors of the Health and Safety Executive.

Kitchens, Staff Restaurants and Canteens

The main fire risk is in the kitchen, which should therefore be separated from the restaurant or canteen by fire resisting construction (half-hour period).

The kitchen should have at least one escape route which is independent of access to the restaurant or canteen. Equally, an escape route from the restaurant or canteen should not pass through the kitchen.

THE REMAINDER OF THE CODE

It should be noted that the Code contains other sections, notably "Fire protection facilities", which covers alarms and detection systems, fire extinguishing systems like sprinklers and fire mains, methods of smoke control and the provision of fire lifts and firemen's lobbies. None of these parts is subject to the Mandatory Rules for Means of Escape but will be important from the points of view of fire certification and building insurance.

APPENDIX

Relevant clauses of BS 5588: Part 2: 1985

The following clauses of BS 5588: Part 2: 1985 are cited in para. 3.3 of Section 3 of the Mandatory Rules. They are listed here together with an indication of the subject dealt with by each clause.

SECTION 1 General	
2 Definitions	
SECTION 2 Analysis of the problem	
4 Planning in relation to fire	1 Spread of Fire
	2 Smoke
	3 Site planning
	4 Mixed user developments
	5 Shops in different occupation
	6 Internal subdivision compartments open vertical open horizontal arcade mall
	7 Ancillary and high risk areas
5 Escape from fire	
	1 General
	2 Evacuation
	3 Disabled persons
SECTION 3 Planning escape from public areas	
6 General	
7 Within and from a storey	1 Distances and number of routes including Tables of distances Travel and direct distance Angle between alternative routes
	2 Widths Table of widths Occupancy density
	3 Escape across a flat roof

8	Stairways and final exits	1	Accommodation stairways and escalators
		2	Number of protected stairways
		3	Siting
		4	Width
		5	Enclosure
		6	Basement stairways
		7	Access lobbies to stairways
		8	External escape stairways
		9	Discharge from stairways and final exits
		10	Examples
9	Small Shops	1	Commentary
		2	Recommendations
			General
			Construction
			Distance of travel and number of escape routes
			Stairways
SECTION 4	Construction, fire safety signs and notices		
10	Construction	1	General
		2	Structural members
		3	Fire resistance
		4	Shafts, lifts, hoists and ducts
		5	Glazed elements
		6	Doors
		7	Doors on escape routes
		8	Door fastenings
		9	Construction of escape routes
		10	Stairway construction
		11	Ladders
11	Fire safety signs and notices	1	Commentary
		2	Recommendations
SECTION 5	Engineering services and ancillary accommodation		
12	General	1	List of engineering services
		2	List of ancillary accommodation
		3	Enclosure of services
		4	Ancillary accommodation
		5	Corridors
		6	Service installation rooms
		7	Higher fire risk areas
		8	Walk-in refrigerated rooms
13	Electrical services	2	Transformer and switchgear rooms
		3	Protected circuits
14	Lighting	5	Lighting of escape routes
15	Heating and fuel storage	1	General
		2	Central heating (water)
		3	Warm air heating
		4	Electric and gas heaters
		5	Electrical thermal storage
		6	Installation
		7	Boiler rooms
		8	Fuel storage
		9	Rooms housing fixed i.c. engines

16	Mechanical ventilation air conditioning and pressurisation	1	Ventilation and air conditioning
		2	Pressurisation
17	Lifts	1	Use of lifts
		2	Lift motor rooms
18	Waste storage and treatment, and incinerators	1	Waste storage and treatment
		2	Incinerators
19	Main storage areas (including receiving and dispatch areas) and covered loading bays	1	Main storage areas
		2	Covered loading bays
20	Car parks	1	Commentary
		2	Recommendations
21	Repair or maintenance workshops		
22	Kitchens, staff restaurants and canteens	1	Commentary
		2	Recommendations

MANDATORY RULES: SECTION 4

Offices

CONTENTS

Offices

INTRODUCTION

The Requirement

The requirement is that –

> ". . . there shall be means of escape in case of fire from the building to a place of safety outside the building capable of being safely and effectively used at all material times."

Satisfying the Requirement

The requirement can only be satisfied by complying with the Mandatory Rules and they in turn require that the building shall be constructed in accordance with the recommendations of certain named sections of British Standard 5588: Part 3: 1983. The named sections are listed in Appendix 4.1. Throughout these notes this BS is referred to as "the Code".

Application

Paragraph B1 in Part B of Schedule 1 to the regulations, in the section headed "Limits on application" states in relation to offices –

> "1 This requirement applies only to:
> (a) a building which is erected and which . . .
> (iii) is or contains an office, . . ."
>
> "2 The means of escape provided need only, . . . in the case of a building containing . . . an office, afford escape for the people from the . . . office."

Appendix A of the Mandatory Rules defines an OFFICE; the definition means that the Rules apply to premises used for any of the following purposes –

1 Administration.
2 Clerical work, including writing, book keeping, sorting papers, filing, typing, duplicating, machine calculating, drawing and the editorial preparation of matter for publication.
3 Handling money as in a bank or building society office.
4 Telephone and telegraph operating.

Paragraph B1 of Schedule 1 clearly applies to a proposal for a new office. In the introductory remarks to these notes on means of escape (page 10), it was held that the Mandatory Rules do not apply to an existing building which is to be extended or materially altered to provide or contain an office. Reasons were given for this interpretation. However, reference was also made to a DoE memorandum which puts forward a contrary view. If the DoE view is accepted, the recommendations of the relevant clauses of the Code will have to be followed.

It should be noted that, under the Fire Precautions Act an office is a "designated building" and even if exempt from the requirement to have a fire certificate under the Fire Precautions (Factories, Offices, Shops and Railway Premises) Order 1989, is subject to the supervisory powers of the fire authority on means of escape and means for fighting fire in accordance with codes of practice issued by the Home Secretary.

Definitions

ACCESS ROOM Provides the only escape route from an inner room.

ACCOMMODATION STAIRWAY
An unprotected stairway connecting two or more storeys; it cannot be considered part of any escape route and may be an escalator.

ANCILLARY ACCOMMODATION
Rooms or parts of the building not used by office staff or visiting public; they are associated with engineering services, storage, plant, computer rooms, etc., which support the main office activity.

CELLULAR PLANNING
A plan form in which storeys are divided into offices of varying size, circulation between them and with the stairway being through corridors. Compare with "open spacial planning" and "open storey planning".

DEAD END An office area or an area of ancillary accommodation from which only one route is available to an exit; or to two exits the routes to which do not diverge by 45° or more and are not separated by fire resisting construction. See also "inner room".

COMPARTMENT
A part of the building separated from all other parts by walls and floors (constructed in accordance with the recommendations of Approved Document B2/3/4, paras. 6.23 to 6.28).

DIRECT DISTANCE
The shortest distance between a point on the floor of a storey and the nearest storey exit, as defined, ignoring all intervening partitions and similar obstacles but remaining at all times within the inner face of the external walls (see Fig. 4.1).

FINAL EXIT The point at which an escape route discharges to a place of safety outside the building, usually at ground level.

FIRE DOOR A door or shutter closing an opening in the protecting structure of a stairway, shaft or duct; also used to control the movement of smoke along a corridor. The door, its frame, door furniture and hinges will together have been certified to resist the passage of fire and/or smoke for a specified period.

HEIGHT OF BUILDING
A measurement taken from the level of the ground or pavement to the highest point of the floor of the top storey. The measurement is made on the centre line of the tallest side of the building. (A top storey which only houses plant is excluded.) Note that this differs from "height" as defined in Approved Document B2/3/4.

HIGH FIRE RISK AREA
An area which, because of the use made of it or materials stored on it, constitutes a higher fire risk than a normal office floor.

INNER ROOM A room from which escape is possible only by passing through an access room.

OPEN SPATIAL PLANNING
A plan form in which floor openings, mezzanine levels or galleries provide vertical interconnection of storeys within the building or a compartment.

OPEN STOREY PLANNING
A plan form for a storey which is the converse of cellular planning. All or most of the storey area consists of one large office the exit from which leads directly into a stairway enclosure or protected lobby.

PRESSURISATION
A method of reducing the penetration of smoke into protected zones by maintaining them at a raised air pressure.

PROTECTED CORRIDOR, LOBBY OR STAIRWAY
A part of an escape route which is separated from the remainder of the building by fire-resisting construction, including fire doors.

STOREY EXIT A doorway between unprotected and protected zones of the escape route. The storey exit usually leads into a stairway (or a stairway protection lobby) or is the final exit.

TRAVEL DISTANCE
The distance which a person escaping from a point on the floor of a storey has to travel, moving around all obstacles (desks, partitions, etc.), to reach the nearest storey exit.

General Considerations

The design and construction of an office building must satisfy all the requirements of Part B of Schedule 1 of the Building Regulations. Requirements B2, B3 and B4 are concerned with structural fire protection and are intended to restrict fire spread within the building. The Approved Document B2/3/4 recommends design solutions which should satisfy the regulation requirements. The principal recommendations are summarised in Table 4.1.

The Code States –

". . . the only sound basis for designing means of escape from fire is to attempt to identify the positions of all possible sources of outbreak of fire and to predict the courses that might be followed by a fire as it develops."

In an office building designed and constructed in line with the recommendations of Approved Document B2/3/4, there should be little risk of fire starting in corridors, lobbies or protected stairways; nor should it start in the structure itself. Fire may result from a fault in the building services, in plant or machinery or in office equipment, or it could start in an office due to human carelessness; failure to extinguish cigarettes or matches properly or the misuse of kettles or electric heaters. Fire may also start in store rooms or kitchens.

Growth and Spread of Fire

In Section 2 of the Code, clause 4.1 contains an explanation of the growth and development of a fire in a building. The important features are that the products of combustion are buoyant and expansive, encouraging outward and upward spread and that it is smoke and toxic gases spreading into and through an escape route which present the main hazard to those making their escape. It is particularly important that the products of combustion and the fire are prevented from reaching ducts, shafts and stairways; in such places they could make rapid progress up the building to break out unexpectedly at storeys well above the storey of origin. The location of fire doors plays a vital part in limiting fire spread; both the Approved Document and the Code make recommendations concerning doors. See later section, "Corridors".

Unless an office building is equipped with an alarm system activated by smoke detectors, smoke will be the first indication that the occupants have of fire in the building. It can cause immediate alarm and occupants will be reluctant to escape through smoke unless they can see or know that a smoke free zone is not far ahead. As the smoke and other gases increasingly contaminate the air a person is breathing, they affect judgement, induce intoxication and, eventually, unconsciousness.

It is important that protected parts of an escape route be kept free of smoke by the proper use of fire doors. In the unprotected areas, the distance a person should have to travel through a possibly smoke-logged corridor or space must be strictly limited. Wherever possible, alternative routes, leading in significantly different directions, should be available; thus enabling a person, wherever he or she may be, to move away from fire and the advancing smoke. The design recommendations in the Code are aimed at achieving these ends.

Form of Recommendations in the Code

In the event of fire, everyone within the building should be able to escape without having to rely on outside assistance. In a new building there should be no need to resort to escape routes across roofs (except where the final exit discharges to a podium roof and the route leads across it and downwards to ground level). The stairways which form part of the escape route should be contained within the building. External fire escape stairways should not be considered for new buildings.

The Code considers the escape route in two parts. Firstly, the horizontal part of the route from any point on a storey to an exit from that storey; the provision of an alternative route comes within this part. In the case of the ground storey, this part of the escape route ends at a final exit.

The second part of the escape route is the vertical part, comprising the stairways which lead downwards and onwards to a final exit to safety outside the building. Lifts are never normally considered as part of the escape route although one or more may be designated for use by chairborne disabled persons.

Code recommendations are given in terms of:

1. The distance from any position on a storey to an exit leading to a protected zone. In some circumstances the distance will be a direct distance; in other cases it will be the travel distance.
2. The availability of alternative routes leading to different exits, the two routes following significantly different directions, i.e. 45° or more.
3. The number of people likely to be occupying particular rooms or areas or needing to use stairways.
4. The number of exits from a room or storey and the number of protected stairways.
5. The width of exits and stairways individually and in aggregate.
6. The siting of exits and stairways in relation to one another.
7. The capacity of inner rooms.

The travel distances, number of exits, floor area limits and exit widths recommended in the Code are aimed to achieve the orderly evacuation of an area, storey, or compartment in 2.5 minutes. For exit widths of 1100mm or more, it is assumed that a unit width of exit of 500mm is sufficient to allow 40 persons per minute to pass through it. The number of people likely to be occupying a room or area may change substantially from time to time or, in some cases, with a change of tenant. In designing the means of escape an average figure, based on observation and judgement, is used; it is determined by dividing room or storey area by a floor space factor. The Code gives factors for rooms used for different purposes. In a room with fixed seating a number based on the number of seats provided is used.

PLANNING

Site Planning

The siting of a single building consisting entirely of office accommodation will be subject to compliance with requirement B4 of Schedule 1 of the regulations. This requirement is intended to limit the risk of fire spreading from one building to another. The side of a building which has openings in the external wall must be positioned sufficiently far from the site boundary to make it unlikely that radiated heat will ignite a neighbouring building (see Approved Document B2/3/4). This influences the positioning of the building on the site and the amount of open space in front of each elevation.

It is possible that there will be fire authority requirements for access to the site for fire appliances, including the provision of areas of hard-standing. In making these requirements they will be using powers given them under the Fire Precautions Act 1971.

The Code draws attention to the need to ensure that, where a final exit discharges to the area outside the building, people should be able to move freely away from the building to a place of safety. The route must not be impeded by, for example, security fencing or vehicles loading or unloading.

An office development may be part of a mixed user complex which contains, in addition to the offices, shops or flats, an assembly building or a hotel. Guidance on fire precautions in such mixed developments has been published by HMSO as Fire Prevention Guide No. 1; Fire Precautions in Town Centre Redevelopment, prepared by the Home Office and the Scottish Home and Health Department.

The advice in the Code relevant to mixed user developments concerns the integrity of escape routes from different occupancies. Each occupancy should have its own means of escape which does not pass through any other occupancy and cannot be threatened by fire in any of them. Alternatively, the possibility of shared escape routes is permissible provided that under no circumstances can fire in one occupancy served by the route make it unusable by another. In deciding between these alternatives, consideration should be given to the type of hazard presented by each occupancy and any fire alarm or detection arrangements fitted. The risk to occupied units (say, flats or a hotel) will be greatest should fire occur in an unoccupied unit (say, an office or a shop) in the same block, particularly if an escape route is shared. The alarm system should alert all occupancies to fire in any one of them.

Internal Planning

Requirement B3 of Schedule 1 of the regulations is concerned with limiting internal fire spread by structural means. Table 6.1 of the Approved Document B2/3/4 relates height, floor area and cubic capacity to the period of fire resistance of structural elements. These limits have a major effect on internal planning and the relevant part of Table 6.1 is reproduced below (Table 4.1).

The minimum period of fire resistance applies to load bearing elements of structure, to elements forming a compartment or providing separation between buildings or parts of a building and to enclosing elements of shafts, protected corridors and protected stairways. The Code also recommends periods of fire resistance for walls and floors enclosing certain high risk areas of ancillary accommodation (see Table 4.7).

Further to ensure the integrity of these elements, Approved Document B2/3/4 recommends the use of "materials of limited combustibility" in their construction. Recommendations will also be found in the Approved Document for surfacing materials for walls and ceilings to restrict fire spread. This is of particular importance in the protected zones of escape routes.

Table 4.1 Offices – size and fire resistance

	MAXIMUM DIMENSIONS			MINIMUM PERIOD OF FIRE RESISTANCE (hr)	
TYPE OF BUILDING	Ht (m)	Max. floor area (m^2)	Capacity (m^3)	Above ground	Basement and floor over
Single storey, or separated part (no basement)	NL NL	3000 NL	NL NL	** { ½ * 1	n.a. n.a.
Building or part which is not single storey	7.5 7.5 15 28 NL	250 500 NL 5000 NL	NL NL 3500 14000 NL	½ * ½ * 1 1 1½	1† 1 1 1½ 2

NOTES

NL = no limit n.a. = not applicable

* In these cases the minimum period for separating walls is 1 hour.

** Fire resistance is only required in external, separating and compartment walls and elements which support them and a gallery.

† If the basement area is $50m^2$ or less, may be half-hour except for separating walls.

Height is measured from mean ground level to mean roof level (not as in the Code).
Floor area is the area contained within the enclosing walls of each storey or compartment.
Cubic capacity is the volume contained within the external envelope of the building or compartment.
(For "Rules of Measurement" see Approved Document B2/3/4; Appendix K.)

The Code recommendations differ according to internal planning arrangements. Two basic floor plans within a storey are considered. The first is "cellular planning" in which the storey is divided into individual offices. The second is "open storey planning", which has very little subdivision of the storey so that office staff work in one large office. It is usual for storeys to be served by two or more stairways, but the Code also makes recommendations for planning means of escape from storeys served by a single, central core stairway.

Internal planning, particularly in open plan offices, is governed by the maximum distance an occupant should have to travel across an open floor area to reach the nearest exit from the storey or doors leading into a protected escape route. Enclosed offices (cellular planning) usually have only one door; the distance from a point in the room most remote from the door, through the door to a point in the corridor where alternative routes become available, is strictly limited. This in turn limits the size of such offices. More information about maximum direct and travel distances will be found in a later section.

Another aspect of internal planning with implications for means of escape is the internal arrangement of spaces and the degree of interconnection of spaces. Two forms of internal planning are significant when considering means of escape provisions: They are –

1. Compartmenting: A building may have every floor constructed as a compartment floor with no interconnection between storeys other than by way of protected shafts and protected stairways. If a fire should break out on one storey, only the occupants of that storey are initially at risk and the means of escape may be designed on the basis that only the occupants of that storey and the storey immediately above it will need to be evacuated in the early stages of the fire.

2. Open Spatial Planning: A building may be entirely without compartments so that all storeys interconnect in some way. The building may consist of split-level floors or floors may be arranged in the form of galleries around a central well. Alternatively, there may be open accommodation stairways running between two or more storeys to allow for the easy circulation of staff. In all these cases the escape route must have sufficient capacity to cope with the simultaneous evacuation of all interconnected storeys.

It is also possible to have a mixture of the extremes of a fully compartmented building and one with all storeys in the building interconnected. In this case some floors, at intervals up the building, may be constructed as compartment floors so that each compartment is several storeys deep. Within each compartment, storeys are spatially open. The means of escape is designed to cater for the simultaneous evacuation of all storeys within a compartment, dealing with each compartment in turn. But it is not necessary to consider the simultaneous emptying of all storeys.

DESIGN OF ESCAPE ROUTES

The Horizontal Part of the Escape Route

In making recommendations for the design of the horizontal part of the escape route the Code distinguishes between occupants in offices and those in ancillary accommodation. A different approach is needed when dealing with a storey laid out as an open plan office and one which is divided into numerous offices of varying size (cellular planning), with a corridor or corridors connecting them. Further precautions will be necessary when several storeys are interconnected (open spatial planning). Special attention should also be given to the provision of satisfactory means of escape for the chairbound disabled. In all cases it will be found necessary to take into account the following interrelated aspects –

(a) the number of people to be evacuated,
(b) the maximum distance a person should have to travel to reach an exit,
(c) the number of escape routes to be provided,
(d) the provision of alternative routes to separate exits,
(e) the width of each exit and the aggregate width of all exits,
(f) the minimum number of storey exits,
(g) escape across a flat roof,
(h) escape from inner rooms,
(i) protection of corridors.

(a) Number of People: In the case of rooms with seating accommodation, for example a restaurant or small conference room, the number of occupants will be known and exit width designed for that number. Elsewhere, numbers are calculated using an occupancy factor which is related to the use of the area in question. Table 4.2 gives the factors recommended by the Code.

Table 4.2 Floor space factors

Description of room or storey	Floor space per person excluding stairway enclosures, lifts and sanitary accommodation (m^2)
OFFICES	
Office	7.0
Open plan office	5.0
Office occupied at minimum area per person permitted by law*	4.0
COMMUNAL ROOMS	
Restaurant, lounge, coffee lounge, dining, meeting or staff room, committee, conference or common room, reading room, waiting room.	1.0
Library, kitchen	7.0
Storage accommodation, car park†	30.0

* Offices, Shops & Railway Premises Act 1963
† Alternatively, 2 persons per car parking space

(b) Maximum Permitted Distances: The Code recommends the use of two distance limits when planning the escape from a storey. The first is the distance a person would actually have to travel from any point on the floor to the nearest exit, moving around all obstacles such as walls, screens and partitions, using openings to reach the storey exit. The Code calls this "the travel distance" and recommends a maximum value.

At the design stage, when building regulation approval is sought, the position of partitions, etc., may not be known; in speculative office building, floors may be "open planned" until they are let. To allow for these circumstances the Code uses a second measure. This is the direct distance from a point on the open office floor to an exit and is referred to as "the direct distance". Figures 4.1, 4.2 and 4.3 illustrate the two distance concepts in storeys with different floor plan arrangements.

The fire authority, when they inspect the completed building for the purpose of certifying fire safety, will measure actual travel distances. If a designer has used direct distances, actual travel distance must be checked when the position of partitions has been decided. It may be useful to remember that "maximum travel distance" is 1.5 × "maximum direct distance".

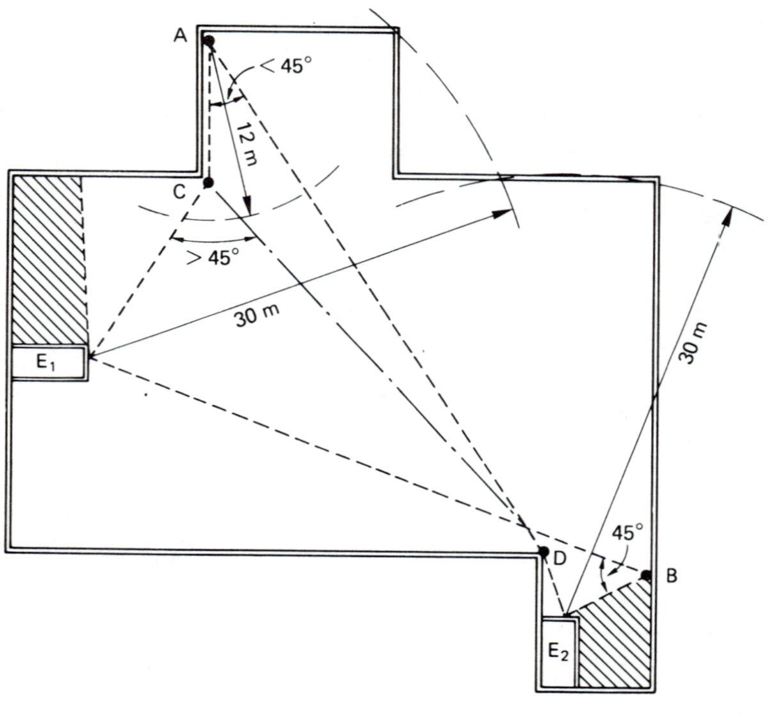

Notes

Two points on the floor, A and B, have been selected.

The lines of direct distance between A and B and the two exits, E_1 and E_2, are shown as broken lines.

The nearest exit to A is E_1 and the route passes through C. Distance ACE_1 must not be greater than 30m. Because angle CAD is less than 45°, distance AC must not exceed 12m. Angle E_1CD is greater than 45°. The 12m radius arc from A is shown.

From B the routes to the two exits diverge by 45°. B is well within the 30m limit from exit E_2.

Arcs at 30m from each exit have been drawn; they indicate that no point on the floor is further than 30m from an exit.

Two shaded areas are shown; from any point within them lines drawn to the two exits enclose an angle of less than 45°. Thus, no point within these areas may be further than 12m from the nearest exit.

FIG. 4.1 Direct distances: open plan office

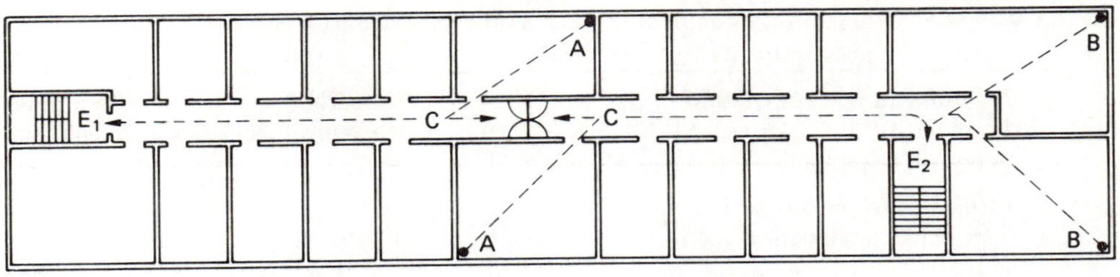

Notes
E_1 and E_2 are storey exits.
Doors (not shown) to stairway enclosures should be FD 30S fire doors.
Doors across corridor between alternative exits should be FD20S.
AC and BE_2 are escape routes in one direction only; max. distance 18m.
From points C alternative routes are available in directions 180° apart.
The maximum distance for AE_1 and AE_2 is 45m.

FIG. 4.2 Travel distances: cellular plan offices

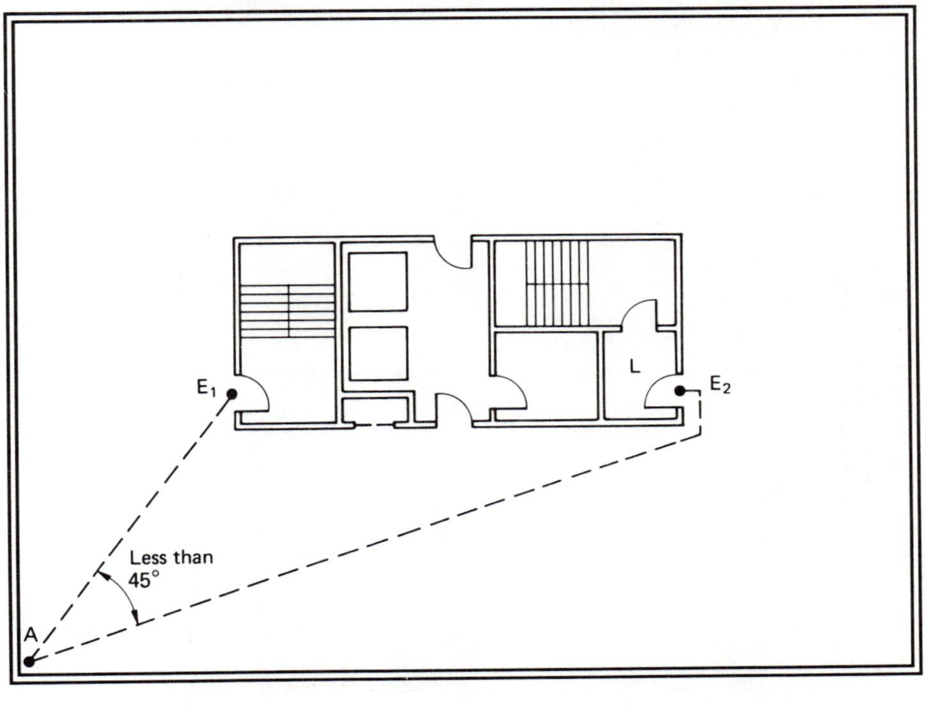

Notes
From any of the corners of the storey, the lines which define the direct distance to the two exits make an angle of less than 45°. The two exits count as one. Consider point A: the nearest exit is E_1 and the distance AE_1 should not exceed 12m.

FIG. 4.3 Direct distance: open plan, central core office

The Code recommends two values for direct and travel distances; the lower figure (shorter distance) applies when travel is possible in only one direction – from an office or inner room or in a dead-end situation. The higher figure is to be used when two or more exits are available. The routes to two exits are either in directions which diverge enough to be considered to offer "alternative means of escape" or are separated by fire resisting construction (see Fig. 4.5).

The Code recommendations for maximum travel distance and maximum direct distance from various situations on a storey to an exit are given in Table 4.3.

Table 4.3 Maximum recommended travel distances in a storey†
(Refer to Figs. 4.1 and 4.2)

Situation and escape option	Maximum travel distance (m)	Maximum direct distance (m)
From a position from which escape is possible in only one direction‡	18	12
Where two routes lead to different exits but are separated by less than 45° and have no fire-resisting construction between them‡	18	12
Where two routes lead to different exits, are less than 45° apart but are separated by a fire-resisting wall for 60% of the distance	45 *	30 **
Where two routes lead to different exits and are separated by more than 45°	45 *	30 **

NOTES

* The first 18m may be unidirectional: (less than 45° separation).
** The first 12m may be unidirectional.
All distances are measured to the nearest exit.
† These distances do not apply in ancillary accommodation involving a high fire risk (see Table 4.7).
‡ These distances cannot be used in offices with open spatial planning in which not less than two adequately separated routes must be provided.

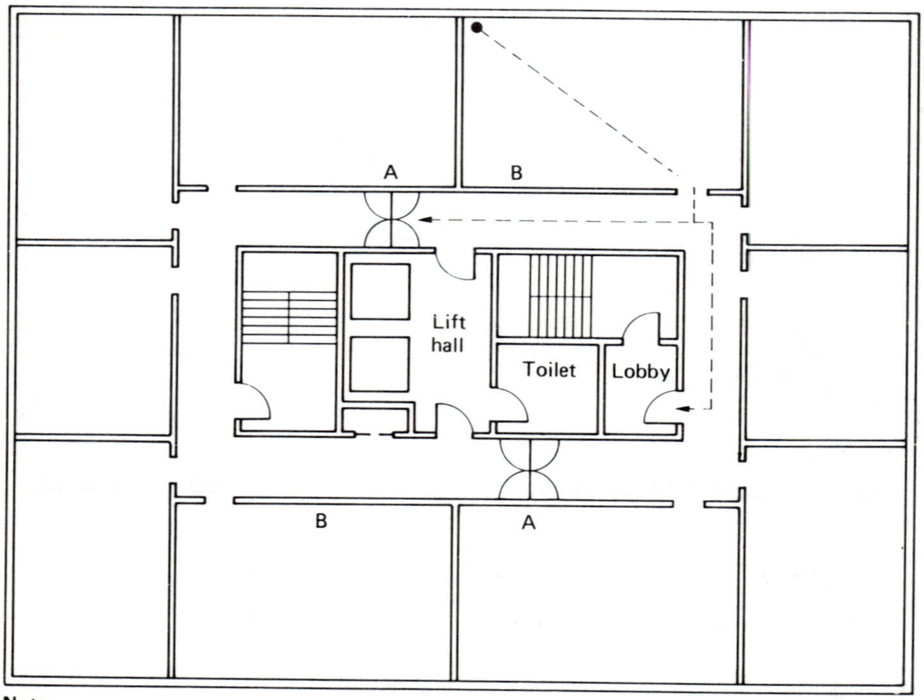

Notes
The access to the two protected stairways is sited at opposite ends of the central core; protected stairways are not to be entered from the lift hall.

The corridor is divided into two parts by fire doors (FD 20S), so providing alternative escape routes from all offices; one route is shown by a broken line.

Corridor fire doors are shown in positions A, A; they could equally well be at B, B but not at A, B. However, if there were no doors between the lift hall and the corridor, doors across the corridor would be recommended in both positions A, A and B, B.

Office doors are not shown but should be fitted. The Code makes no recommendation for fire resistance or self-closing devices for office doors.

One stairway is shown approached through a protected lobby; such a lobby would be necessary for both stairways if the building exceeded 18m in height. For buildings less than 18m high, neither stairway needs a lobby approach.

FIG. 4.4 Central core, cellular plan offices

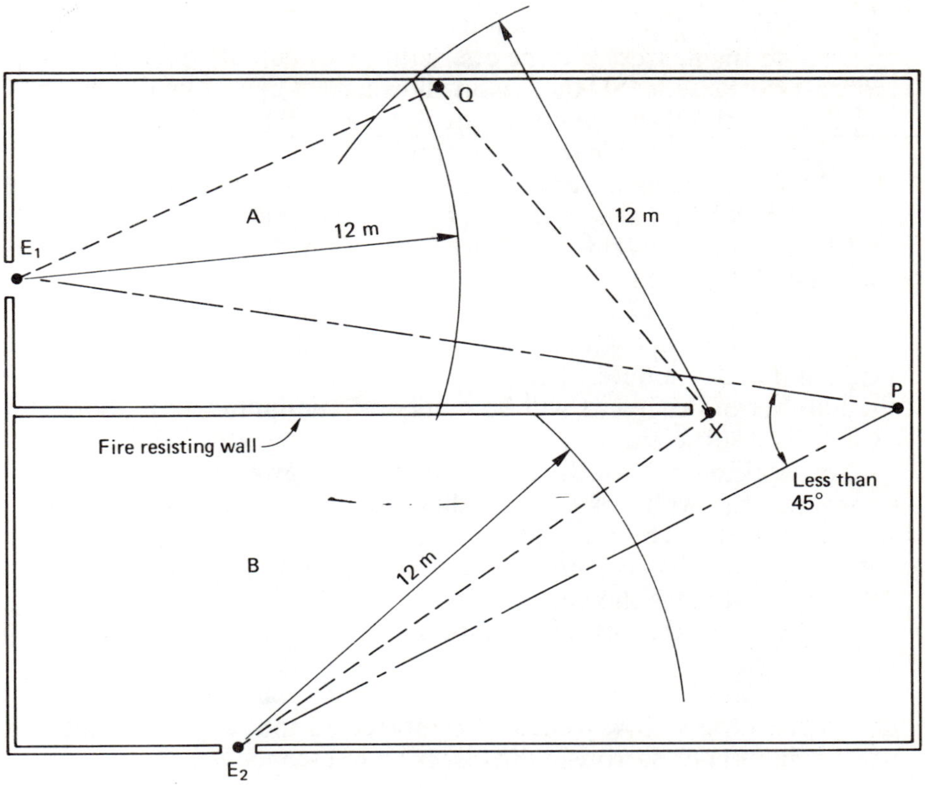

Notes
From point P the angle between the routes to E_1 and E_2 respectively is less than 45°, but the fire resisting wall makes the routes acceptable as "alternative routes".

From point Q the shortest route leads to E_1; the alternative, QXE_2 has a total length of less than 30m and the portion QX is less than 12m.

Note the arcs of circles: From E_1 a 12m arc describes an area A from which a single escape route to E_1 is acceptable; A similar 12m arc has been drawn with E_2 as centre defining area B.

FIG. 4.5 Fire-resisting construction separating escape routes

(c) Number of Escape Routes: The general rule is that there should be at least two escape routes available from any storey or room. With open interconnection between storeys (open spatial planning) it is essential to provide at least two escape routes from every storey or working level. There are, however some exceptions to the rule.

Table 4.3 recommends two values for both travel and direct distance; provided the shorter distances are adhered to, a room may have only one escape route. The same applies to an escape route from an inner room.

If a room is not likely to be used by more than 50 people, it too can be planned with only one escape route, An entire storey may be served by one escape route if it is in a building of not more than three storeys above ground or not more than 11m high.

(d) Alternative Routes: Normally, a person escaping from a fire will follow the shortest route to a storey exit. But if that shortest route should be made unusable because of fire or smoke, except in certain clearly defined circumstances, the person should be able to follow an alternative route, turning away from the fire.

In a cellular plan office, unless it is a very large one, the first part of the escape route will lead to a single exit (the office door) with no alternative route available. Once in the corridor, in buildings with two or more stairways, alternative routes separated by 180° will be available.

In an open plan office there is a first route to the nearest exit; an alternative route should also be available offering escape in a significantly different direction.

"Significantly different" is defined in Fig. 4.1. The lines representing direct distance to the nearest and the alternative exits should enclose an angle of 45° or more. It will be seen that there is an allowance of up to 12m which can be at a lesser angle. Note that achieving this angular separation makes it necessary to locate storey exits at a distance from one another.

If two exits from a storey are so close that the angle between the routes is less than 45° they count as one unless, for a proportion (60 per cent) of the distance, they are separated by fire resisting construction. The longer is "alternative" to the shorter (see Fig. 4.5).

(e) Width: The width of an exit controls the number of people who can pass through it in 2.5 minutes. In offices, particularly where a cellular plan is adopted, the number of occupants will be small and minimum width doors, corridors and exits will be sufficient.

In working out width requirements for a given population, two values are important; the width of an individual exit and the combined widths of all exits serving a storey. The relationship between exit width and the number of people able to pass through it in the standard escape time is shown grapically in Fig. 4.6. The Code provides this information in a table with widths increasing in 100mm steps; the Code accepts interpolation and extrapolation and a graph makes this process easier.

The width of a corridor at its narrowest should not be less than that of the doorway to the stairway enclosure, protected lobby or final exit. Escape route widths should not narrow in the direction of escape.

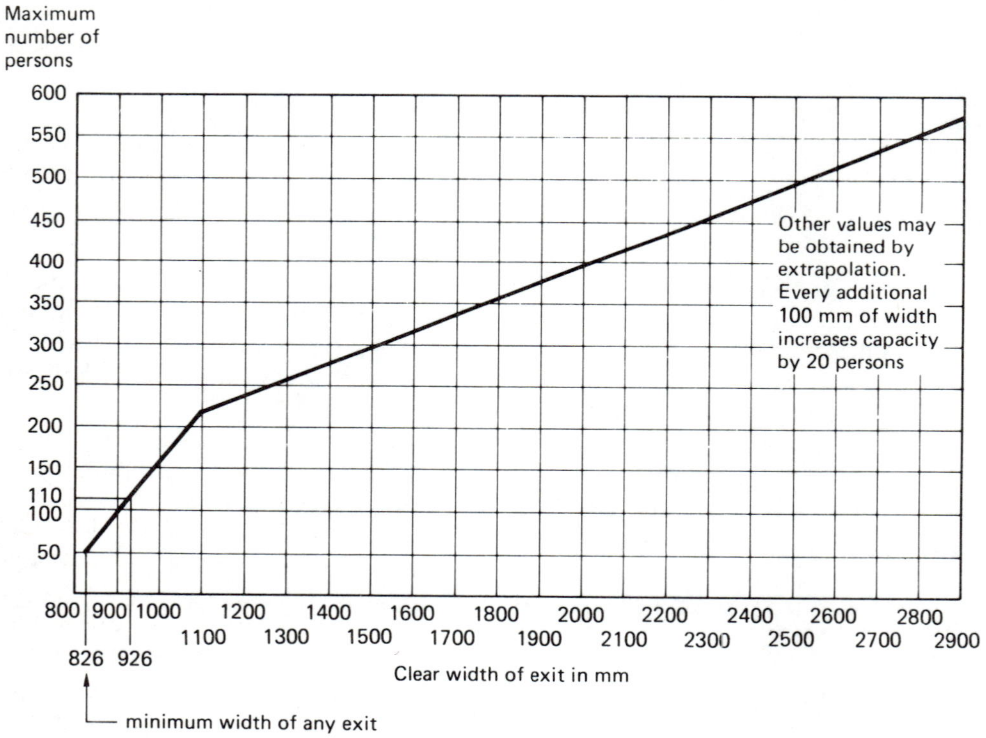

FIG. 4.6 Capacity of storey exits (persons per 2.5 min)

(f) Minimum Number of Storey Exits: The number of storey exits to be provided will be determined by several factors. First there is the need to provide main and alternative escape routes, at adequate angular separation, from all parts of the storey; then a sufficient aggregate width is needed. It is also a Code recommendation that each exit in turn must be assumed to have been rendered unusable because of fire.

(g) Escape by Way of a Flat Roof: When designing a new building the use of a flat roof as part of an escape route should, preferably, be avoided. When a building stands on a podium it is better if the final exit is at ground level rather than have it discharge on to the podium roof; if discharge to the podium roof is unavoidable, it is essential to ensure that a clearly marked route, guarded where necessary, leads onwards away from the building to safety at ground level.

If there are unusual circumstances which give a designer no choice but to use a flat roof for part of an escape route, the Code makes the following recommendations:

- There should be an alternative route available from the area served by the roof escape route.
- The roof must be part of the building served by the escape route.
- The route should lead to an exit on another storey.
- The route should be adequately marked and guarded in accordance with BS 6180: 1982: "Code of Practice for protective barriers in and about buildings".
- That part of the roof forming the escape route should be constructed as a fire resisting floor and the structure which supports it should also have appropriate fire resistance.
- The route should not pass closer than 3m to ventilators, non fire-resisting roof lights, windows, doors or any other opening through which fire or smoke could discharge to make the route unusable.

(h) Inner Rooms: Inner rooms should not be designed to accommodate more than 50 people. Escape from an inner room will, of necessity, be through an access room (see definition). The travel distance limit appropriate to a "dead-end" applies and the route is measured from a point in the inner room most remote from the door, through the access room, to a doorway leading to a protected zone. The access room should not be a room containing a high fire risk.

Arrangements should be made for a person in an inner room to receive early warning of fire in the access room. This can be done either by fitting a suitable alarm system operated by a smoke detector in the access room, or by providing a clear view of this room through glazing in the door or wall (or both). As an alternative to glazing, the partition between the inner room and the access room could be stopped well short of the ceiling.

(i) Corridors: Corridors which form part of an escape route are of two types. First, there is the corridor which connects two protected stairways and which may be used in either direction. Or there is the corridor serving a dead-end. The latter type needs to be both more secure and shorter since it may be necessary for the occupant of a room served by it to pass a room which is on fire. There is no alternative means of escape.

In both types of corridors, partitions should be carried up to ceiling level and a fire-tight joint provided between partition and ceiling. If there is a false, suspended ceiling, the partitions should go through it to meet the structural floor above. All openings (except those serving an enquiry or reception office) should be fitted with doors or shutters. The Code does not recommend a period of fire resistance for room doors, so unless the room contains a high fire risk, the doors need not be fire doors.

In dead-end corridors, the partitions should, in addition, be fire resisting (minimum half-hour) and room doors should be FD 20S fire doors and self closing. The length of a dead-end corridor will be restricted by travel distance limitations.

Corridors longer than 12m which connect two stairways, linking corridors and dead-end corridors should have fire doors (FD 20S) across their width in certain positions. The Code gives no positive advice about locating these doors, but refers the reader to a series of floor plans which, unfortunately, have no scale indicated.

As shown in Fig. 4.2, a length of corridor connecting two stairways should have a set of doors at about its mid point. Doors should also be placed at the end of a dead-end corridor to prevent smoke from a fire anywhere in the dead-end from spreading into the corridor which runs between the two stairways (see Fig. 4.7). Figure 4.7 also shows a dead-end corridor which joins a corridor between two stairways at about its mid point. Two sets of doors are placed across the main corridor, one each side of the junction; they ensure that, wherever a fire occurs, only one "leg" of corridor can be affected. Figure 4.4. shows how doors should be positioned in the corridor around a central core.

Travel distance is another factor which affects the location of doors in a long corridor. The total distance from the innermost point in any office to a set of doors should not exceed 45m when escape from just outside the office is available in alternative directions.

Dimensional limits apply to recesses off corridors and short extensions of corridor beyond protected stairways; these are set out in Figs. 4.8 and 4.9.

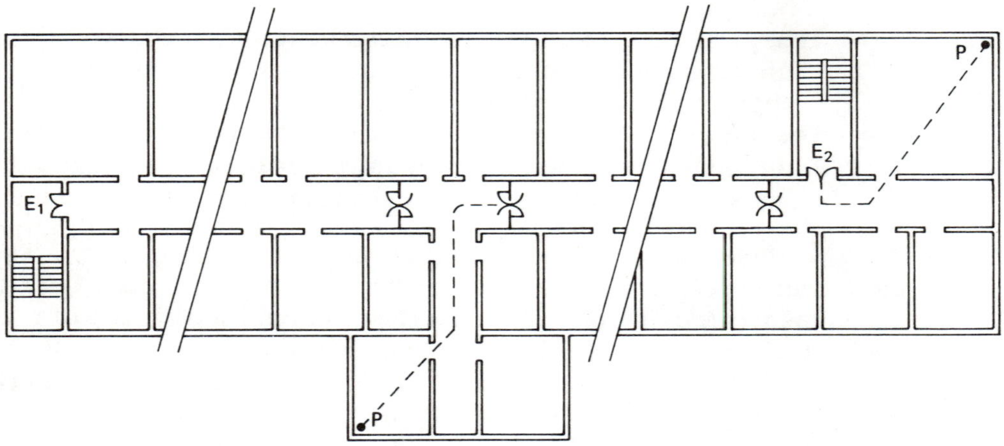

Notes

Double-swing doors are located across the corridor so as to separate dead-end areas from lengths of corridor which offer escape in two directions 180° apart.

Within each of the dead-end areas the maximum permitted travel distance from point P to the nearest set of doors may not exceed 18m. Nowhere in any office shall be more than 45m from one of the storey exits E_1 or E_2.

FIG. 4.7 Fire doors in corridors

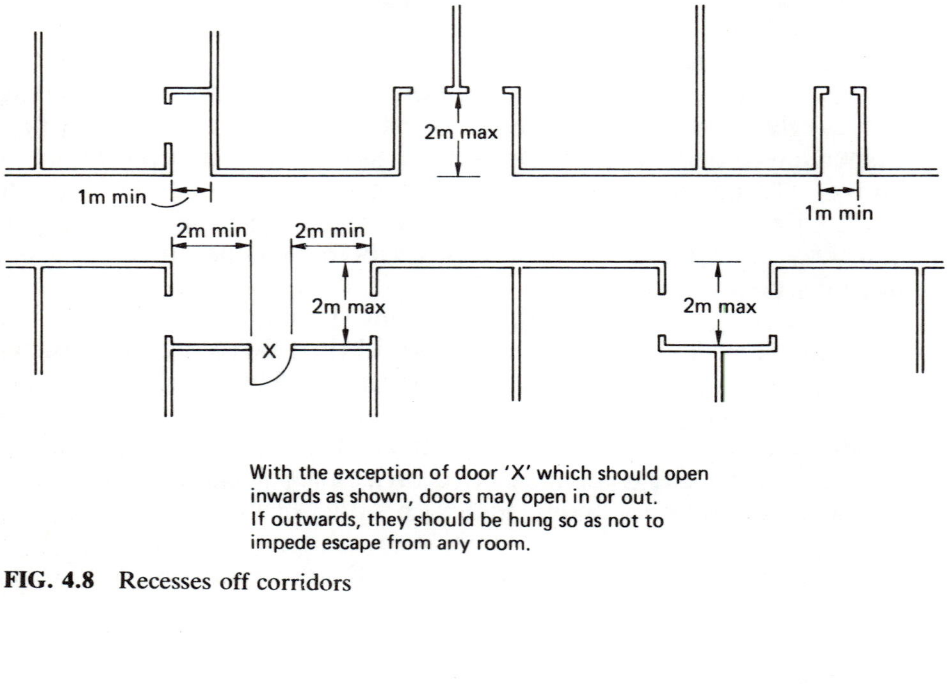

With the exception of door 'X' which should open inwards as shown, doors may open in or out. If outwards, they should be hung so as not to impede escape from any room.

FIG. 4.8 Recesses off corridors

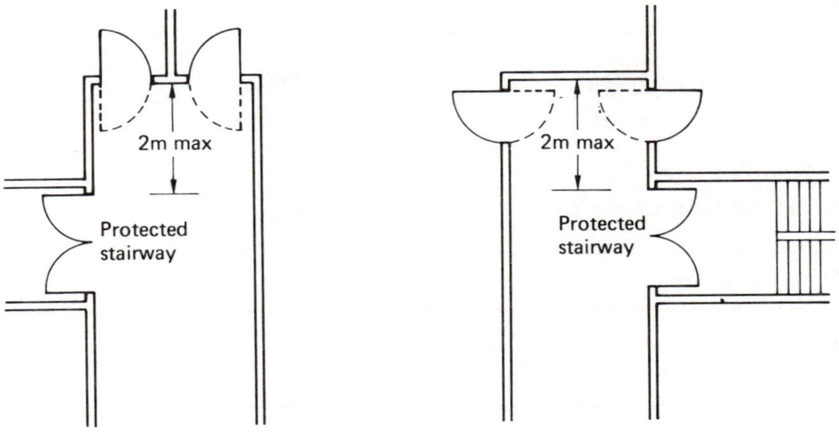

- only two rooms may open off the 2m extension
- room doors may swing in or out

FIG. 4.9 Corridor extending beyond a protected stairway (short dead-end)

The Vertical Part of the Escape Route – Stairways
Lifts, escalators and open stairs cannot be considered as part of the means of escape (except that a lift may provide the means of escape for chairborne disabled persons). Only protected stairways can be used for the vertical part of the escape route. Such stairways are enclosed in fire-resisting construction and are separated from the horizontal part of the escape route at each storey by self-closing fire doors and in some circumstances by protected approach lobbies.

Paragraphs 6.29 to 6.34 and 6.42 to 6.44 of Approved Document B2/3/4 give detailed recommendations for structural fire precautions. The enclosure must have adequate fire resistance, Table 4.1, and construction materials should be "materials of limited combustibility". There are restrictions on what the enclosures may contain by way of services, cupboards, etc., and on the location of windows. The main features of protected stairways which contribute to a safe means of escape are–

(a) The siting of stairways.
(b) The number of stairways to be provided.
(c) The width of each stairway and the aggregate width of all stairways.
(d) The enclosure of stairways.
(e) Additional protection for stairways to basements.
(f) Protected lobby access to stairways.
(g) Safe discharge at the foot of stairways.

(a) Siting of Stairways: The siting of stairway enclosures must take account of the exit requirements on all storeys served and the need to afford effective alternative directions of escape in the horizontal part of the route. It is also necessary to ensure that the final exit at the foot of the stairway discharges to a place of safety (see "Site Planning", page 93, and item (g) below). Storey exits giving access to protected stairways should be sited away from lifts and lift halls, the heads of escalators or the top of accommodation stairways.

(b) Number of Stairways: The number of stairways required will depend on whether the building is compartmented at every floor or has some vertical open planning. With compartmenting, stairway capacity is based on the simultaneous discharge of two storeys; with open spatial planning the stairways must be able to cope with the discharge of all storeys together. In determining the number of stairways required, it must always be assumed that one stairway is rendered unusable by the fire. Thus, from all open plan storeys and the larger cellular plan ones, the minimum number is two. In some small offices – those which do not

exceed three storeys above the ground storey, or those in which the floor level of the upper storey is not more than 11m above ground level, and in which, in addition, travel distances appropriate to "dead-ends" are not exceeded – one stairway will be acceptable (see Table 4.4).

Table 4.4 Single stairway building – width of stairway

Maximum number of persons per storey	Width of stairway (mm)
50	900
more than 50	1100

(c) Width of Stairways: The width of a stairway is determined by the number of persons who will be using it in the event of fire. This in turn depends on the population of each storey and on how many storeys will be discharging simultaneously on to the stairway in the early stages of a fire.

The calculation of width also takes account of the fact that a stairway can hold a certain number of people in each of its interstorey zones. In the majority of office buildings the population of each storey will be small and a minimum width stairway (1000mm) will cope with the numbers involved.

Figures 4.10 and 4.11 show the relationship between stairway width and capacity. Note that the width required is the aggregate width of the stairways when one is discounted. Thus with a two stairway building, each stairway must be wide enough to handle the escaping population on its own. Widths do not all have to be the same but none may be less than 1m.

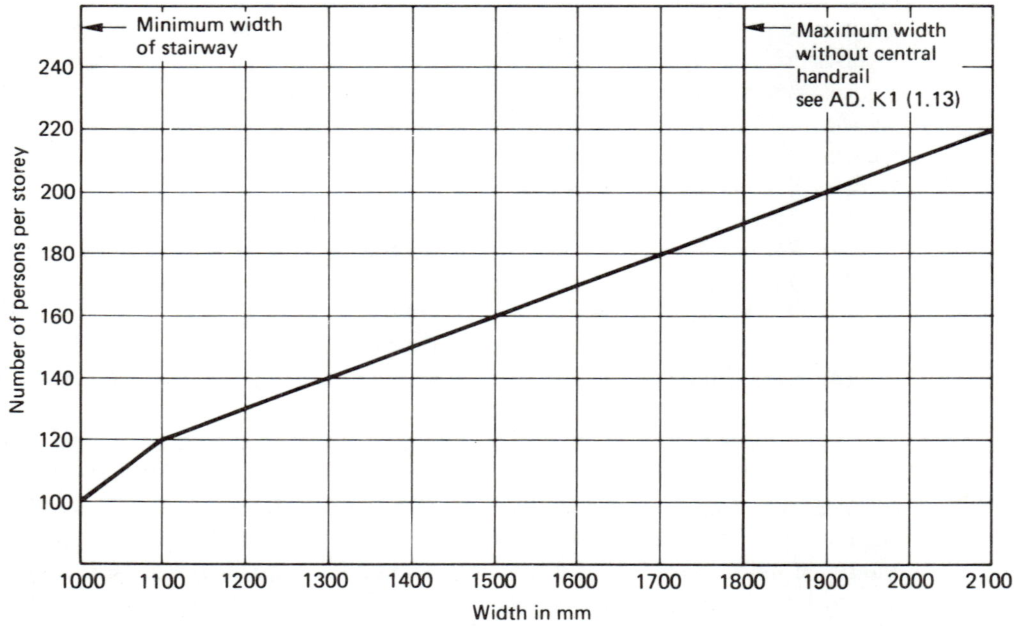

FIG. 4.10 Stairway width for the evacuation of two adjacent storeys in a horizontally compartmented building

(d) Enclosure of Stairways: The fire resistance of the enclosure should be not less than 30 minutes; with buildings which have many storeys or large floor areas, Approved Document B2/3/4 may recommend longer periods (see Table 4.1). The enclosure may contain sanitary accommodation, provided it is not used as a cloakroom and does not contain a portable heating appliance or any gas appliance other than a water heater or incinerator. Cupboards may be located

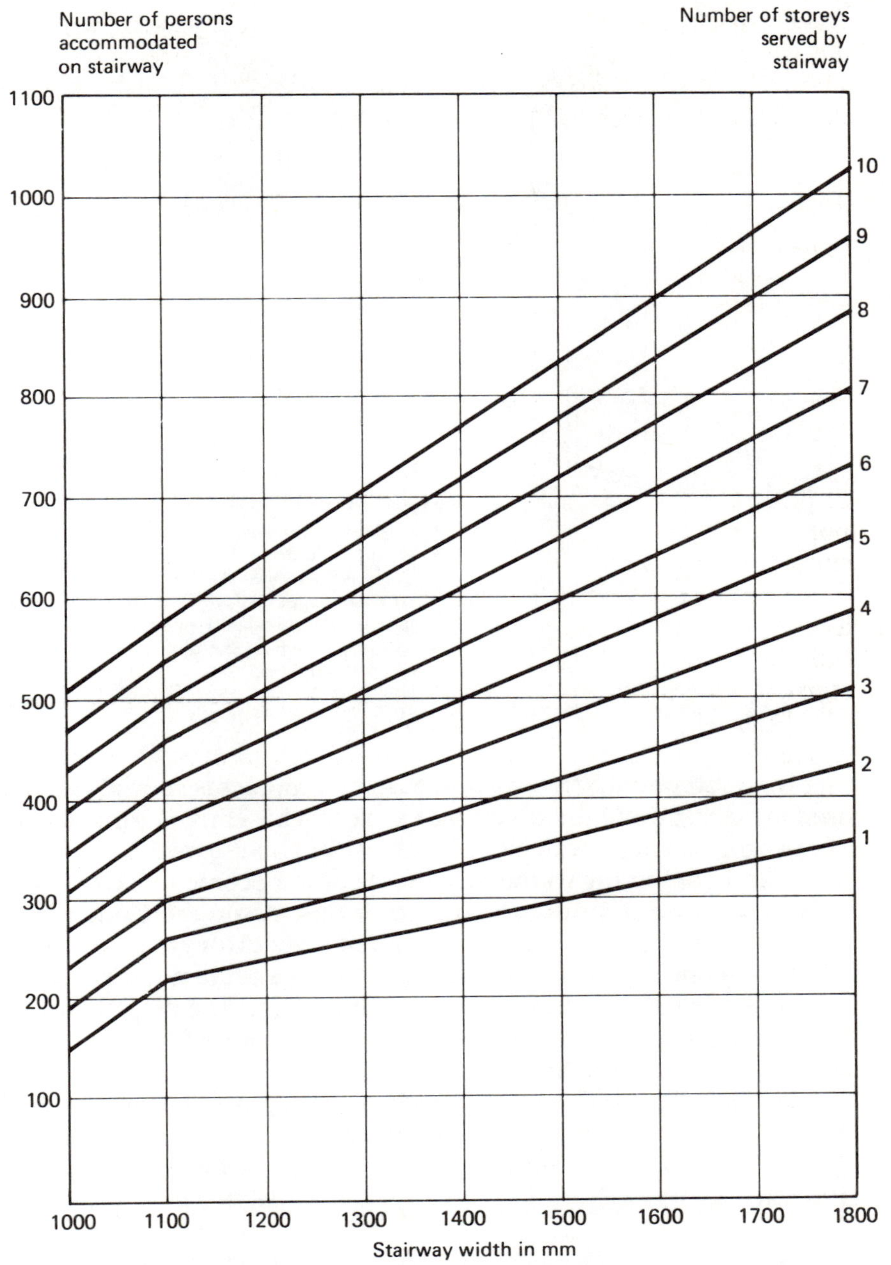

FIG. 4.11 Stairway width for the evacuation of basements and for the total evacuation of the building

within the stairway enclosure provided they themselves are enclosed in fire-resisting construction. Only one stairway should occupy a protected enclosure and there should be only one stairway entry doorway on each storey.

It is important to ensure that a stairway cannot be rendered unusable because of fire or radiated heat penetrating by way of windows. This is unlikely if the outer wall of the stairway is flush with the elevation of the building. But if a protected stairway is recessed behind the line of the elevation or projects in front of it, the threat of fire penetrating the enclosure becomes a real one.

To counteract this threat, a minimum distance of 1.8m should be maintained between windows at the same level in walls at right-angles to one another. If a window in the building is within 9m vertically below a stairway window and within 1.8m horizontally, the former should have fixed fire-resisting glazing. Figure 4.12 illustrates this recommendation.

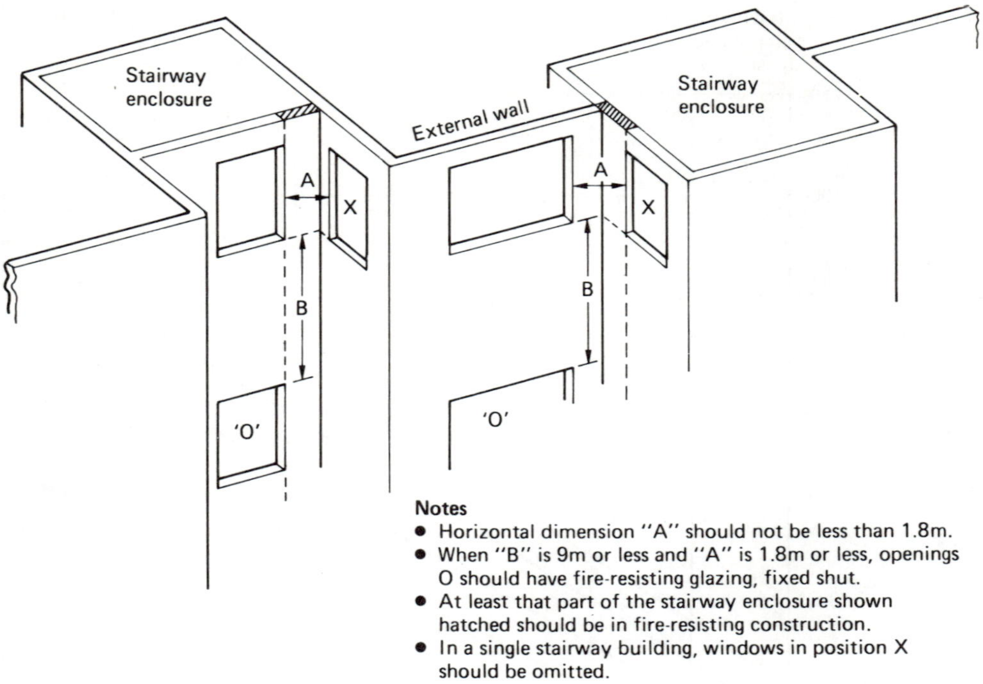

FIG. 4.12 Stairway enclosures – recessed or projecting

(e) Basement Stairways: A basement presents special risks. It is frequently used for storage and the severity of a fire in a basement is increased by the absence of openings through which heat and smoke can escape. A stairway serving upper floors and continuing to the basement could become unusable because of smoke unless other precautions are taken. For this reason the Code recommends that it should be approached at every basement level through a protected lobby.

In a single stairway building the stairway should not serve a basement storey. The basement should have its own stairway, leading directly to the open air and discharging in a position where smoke or fire emitted from it cannot threaten any escape route from the upper storeys.

In an office building in which upper floors are served by two or more stairways, at least one of them should end at ground level. Another stairway may be continued down to the basement provided at every basement level there is a protected lobby.

(f) Protected Lobbies: The occupants of high-rise offices are at greater risk than are those in low-rise buildings. In high buildings (those of height greater than 18m), there should be a protected enclosure separating the stairway from the storey. The only exception is a top storey used exclusively for plant rooms. The protected enclosure may be called a lobby if it is roughly rectangular in shape; or it will be called a protected corridor if it is longer and narrower.

Similar lobby or corridor protection is recommended by the Code in other circumstances. In a single stairway building (not more than three storeys above the ground storey) there should be a lobby at every storey except the top storey. This does not apply, however, if there is only a ground and first storey. If a stairway connects above-ground storeys with basement storeys or an enclosed car park, or if a stairway serves only basement or car parking storeys, it should have protected lobbies at every basement or car park level. Any corridor connecting high risk rooms (boiler room, transformer chamber) with a stairway should be similarly treated.

To summarise, the Code recommends protected lobby approach in the following circumstances:

- Any stairway in a building greater than 18m in height, at every upper storey.
- Any stairway which connects the ground or upper storeys to a basement

storey, or one which serves more than one basement storey, at every basement storey.

- Any stairway giving access to an enclosed car park, at every car park level.
- Any stairway which directly serves a high fire risk area, for example, boiler room or transformer chamber, at the level of that risk.

The fire-resisting enclosure for a protected lobby, the services, etc., permitted within it and the protection of the opening giving access to it are the same as for a protected stairway. There are also recommendations for the lobby to be ventilated in certain situations.

(g) Discharge from Stairways: The escape route ends at a final exit which gives direct access to a place of safety away from any threat from fire in the building, including fire in the ground storey or in a basement. A final exit serving office accommodation should be well separated from any access to an area of high fire risk such as a boiler or transformer room.

A final exit may be at the foot of a stairway which opens directly to the street in the open air. If the street is an enclosed arcade or mall, special arrangements will be needed to control smoke and ensure that those escaping can disperse rapidly to places where they are no longer in danger from the fire.

Stairways serving buildings which have not more than four storeys may discharge into a concourse or courtyard or onto a walkway. In all cases the route onwards should be clearly marked; walkway widths and the exit from a courtyard must be wide enough not to impede the passage of all who may need to use it. (That is, must have a capacity to match the aggregate width of all stairways discharging into the courtyard.)

If an office block is built off a podium, stairways are better carried down through the podium to discharge at ground level. If this is not possible for some reason and one or more stairways ends at podium level, a safe route should be clearly marked across the podium deck to other stairways leading to the ground.

It is accepted that one stairway in a building which has two or more protected stairways may end in an entrance hallway which also contains a reception desk or enquiry office. The reception area or office should not be larger than $10m^2$. (Note that two protected stairways should never discharge into the same ground floor space otherwise a fire in that area could put both out of use.)

If a protected stairway connects with basement storeys, the final exit at ground level should be clearly marked so that those escaping from upper floors do not continue downwards into the basement.

A stairway may connect with an exit passageway which leads to the final exit; if two stairways do this there must be two separate passageways with no connection between them. That means that any common wall between the passageways should be imperforate.

(h) External Escape Stairways: The Code gives advice about external fire escape stairways and the clauses dealing with the matter are among those cited in the Mandatory Rules. In a new office building, external stairs of the exposed, iron or steel type should not be considered as a satisfactory means of escape. This is not to say that offices may not be planned with separated, fully enclosed stairway towers, connected to the building by a bridge at each storey level. Such a stairway should be constructed to conform with all the recommendations made for internal stairways.

CONSTRUCTION

General

Buildings which conform to the requirements B2, B3 and B4 of current Building Regulations will satisfy the recommendations of the Code for fire resistance, etc. However, the Code emphasises two additional points.

If a structural element is made fire resistant by enclosing it in a casing or membrane, such protection can be vulnerable to mechanical or other damage in use, so that the fire-resisting quality is destroyed or reduced. The Code recommends care in the selection of building methods and materials for casings and membranes.

The second point concerns junctions and joints; care should be taken to avoid weaknesses at junctions between elements (for example, between wall and ceiling), or at joints between sheet materials (membranes) which provide fire protection to an element of structure. The Code also recommends care in choosing the method of fixing sheet materials and casings so that they remain in position for the required period when subjected to a fire.

Glazed Elements

Areas of glazing in walls intended to protect an escape route or enclose a stairway can present a hazard to people escaping if they do not prevent heat radiation. Some glazing, when subjected to the standard fire test, can provide resistance in terms of both integrity and insulation for a half-hour period; this is sufficient for means of escape. Such glazing can be used without restriction in parts of the building which are the concern of the Mandatory Rules.

Most glass and glazing does not meet the insulation requirement for more than a few minutes and the Code recommends restrictions in its use. These are as follows:

(1) In buildings or parts which have a single stairway

There should be no glazing in a fire-resisting wall: directly between –

(a) a protected stairway and an area used as an office or a non-fire-resisting corridor.

Any amount of glazing may be used in those parts of a wall which are at least 1.1m above finished floor level:

(b) a protected stairway and protected lobby,
(c) a protected stairway and a fire-resisting corridor,
(d) a protected lobby and office areas,
(e) a fire-resisting corridor serving a "dead-end" and office areas.

In any leaf of a fire door which has been shown to be suitable for glazing, the proportion of glazing may be:

(i) 25 per cent of the area of a leaf of a door in situation (a) above;
(ii) an unlimited area of a door in any of situations (b) to (e).

(2) In buildings or parts with more than one stairway

Any amount of glazing may be used in those parts of a wall which are at least 1.1m above finished floor level in situations (a) and (e).

An unlimited amount may be used in situations (b), (c) and (d) and between a protected corridor not associated with a "dead-end" and the office area.

In any leaf of a fire door which has been shown to be suitable for glazing, the proportion of glazing may be:

(i) 50 per cent of the leaf of a door in a wall in situation (a);
(ii) an unlimited area of the leaf of a door in situations (b) to (e).

Fire Doors

Fire doors are classified by reference to the period in minutes during which their integrity is maintained under test. A door is thus designated, for example FD 20, if integrity is sustained for 20 minutes. Where doors are, in addition, required to retard the passage of smoke at ambient temperature, the letter S is added; this generally means that it will be fitted with a flexible edge seal.

Fire doors along an escape route must be easy to open from the escape side of the door; latches should operate easily without the need of a key. If a door on an escape route must be kept locked for security reasons, the precautions to be taken should be discussed with the fire authority.

Table 4.5 Fire doors

Location of door	Recommended designation
In the enclosure of a protected stairway	FD 30S
In the enclosure of a protected lobby or protected corridor	FD 30S
An office door in a corridor forming a dead-end	FD 20S
In the enclosure of a storage area not greater than 450m^2 which is not a high fire-risk area	FD 30
In the enclosure of workshops and reprographic rooms	FD 30
In the enclosure of kitchens	FD 30
In the enclosure of high risk ancillary accommodation, boiler room, fuel store, car park	FD 60
In lift shafts opening into unprotected areas	FD 30
In openings into ducts, etc.	FD 30
Across corridors more than 12m long which are part of an escape route	FD 20S
Between a dead-end portion of corridor and the remainder of corridor	FD 20S

If the number of occupants who will be using a fire door in the event of fire exceeds 50, the door should be hung so that it opens in the direction of escape. The swing of a door should not reduce the effective width of any escape route; this is especially important with doors opening onto landings which will have a flow of people coming down the stairs from the floor above. Every door should open at least 90°. Any door which opens outwards onto an escape route corridor should be set back in a recess at least as deep as the width of the door. See Fig. 4.8 for more about doors in recesses.

Double swing doors across an escape corridor should either be fully glazed or be provided with a vision panel. Revolving doors or automatic sliding doors should be flanked by normal swing doors of the full width unless they are arranged to fail-safe to outward opening from any position of opening.

All fire doors, except those to cupboards or service ducts, should be self-closing; rising-butt hinges cannot be relied on to provide this function. The self-closing device should not be one which can easily be disconnected or immobilised and it should be powerful enough to override any latches; it should be impossible to retain the doors in the open position except in the manner described in the next paragraph.

A fire door may be in such regular use, for example, by staff with arms full of files or pushing wheeled trolleys, that it is inconvenient not to have it permanently open during working hours. Such a door may be fitted with an electromagnetic or similar device which will hold it open against the self-closing device. The magnetic catch must be fitted so that the door will be released automatically either by a smoke detector system, or a manual alarm system, or by manual operation at a central control or by a power failure.

Cupboard doors or service duct covers which are not required to be self-closing should be kept locked shut when not being used; they should be marked at about eye level on both sides “KEEP LOCKED” (see BS 5499: Pt 1).

Hinges for fire doors should be non-combustible and should not have a melting point below 800°C. Other door furniture should be simple and easily operated from the escape side. Panic bolts should comply with BS 5725: Part 1; any other security device should have fire authority approval.

Construction of Escape Routes

All corridors which form part of an escape route, the landings and treads of protected stairways and entrance halls or corridors at ground level which lead onwards to the final exit, should be constructed so as to provide a safe passage for occupants using them in an emergency. The following features are important:

- The use of non-slip floor finishes.
- Adequate headroom – not less than 2m.
- Freedom from wall projections which could catch on clothing or otherwise impede the flow of people using the route.
- Ramp gradients no steeper than 1 in 12.
- Stairways constructed in accordance with Part K of Schedule 1 of the regulations in respect of going, handrailing, balustrades and headroom. The treads should have a non-slip finish.

ENGINEERING SERVICES AND ANCILLARY ACCOMMODATION

General

Engineering services and ancillary accommodation are important from the point of view of means of escape for two main reasons. First, there is the potential fire hazard some present to building occupants and the escape routes provided for their safety. Second, there is the threat they may present to staff working in the accommodation or directly involved with the service installations.

The Code lists all engineering services but only some have implications for means of escape. They are:

- Protected electrical circuits which serve emergency lighting, alarm systems or other fire-related equipment.
- Emergency lighting of escape routes.
- Boiler rooms and fuel storage.
- Ventilation and air conditioning systems.
- Protected stairway pressurisation systems.
- Lift motor rooms.
- Refuse disposal systems.

Ancillary accommodation in this Code consists of those areas not usually occupied by office employees or visitors to the offices. It consists of rooms associated with engineering services, telephone exchanges, storage areas, refuse facilities, workshops, kitchens, computer rooms and car parks.

B3 of Schedule 1 of the regulations requires a building to be so constructed as to maintain its stability for a reasonable period in the event of fire. Approved Document B2/3/4 interprets that requirement by recommending a period of fire resistance to be achieved generally by structural elements according to the size (height, area, capacity) of the building. The recommendations for offices are reproduced in Table 4.1.

The Code goes further than this in relation to areas of ancillary accommodation. It divides the areas into three groups – roughly, low, medium and high hazard – and recommends a period of fire resistance for enclosing structure (walls, floor, ceiling) for each group. The Code recommendations are reproduced in Table 4.7.

Some engineering services are known to be potential sources of fire, particularly if incorrectly installed or badly serviced. To reduce this potential hazard, the Code recommends that the installation of services be carried out to conform with the appropriate British Standard. Those quoted are reproduced in Table 4.6.

It is an increasingly common practice in offices to provide a platform floor under which a variety of wiring for computer terminals, telephones, etc. can be run to individual work stations. The cavity formed beneath such a floor should be fitted with suitable cavity barriers in accordance with the recommendations of Appendix G of Approved Document B2/3/4. The Code makes reference to *Method of Building MOB 01.801 Performance Specification: Platform Floors* (1981) which is published by the Property Services Agency of DoE.

The Code also recommends that engineering service ducts be constructed to comply with CP 413. Reference to Approved Document B2/3/4, Appendix L, definitions, will indicate that a service duct is one form of protected shaft; Appendix E makes detailed recommendations for the construction of protected shafts.

Table 4.6 British Standards for service installations

Category of installation	British Standard
Lift motor rooms	BS 5655: Pt 1: 1979
Waste storage chambers, chutes and hoppers	BS 5906: 1980
Pressurisation systems	BS 5588: Pt 4: 1978
Mechanical ventilation and air conditioning	BS 5720: 1979
Service and ventilation ducts	BS CP 413: 1973
Oil-fired installations	BS 5410: Pt 1: 1977 Pt 2: 1978
Oil storage tanks	BS 799: Pt 5: 1975
Town and natural gas installations	BS 5376: Pt 2: 1976 BS CP 332: Pt 3: 1970
Escape route lighting	BS 5266: Pt 1: 1975

Ancillary Accommodation – Means of Escape

Some rooms used for ancillary purposes present no greater hazard than offices; examples include computer rooms and telephone exchanges. For them the recommendations for offices are appropriate. The remainder present a greater fire risk, either because they contain some flammable material, or are only irregularly

Table 4.7 Period of fire resistance for structure enclosing ancillary accommodation

Type of accommodation	Minimum period of fire resistance of enclosing walls, floor and ceiling and for doors or shutters to openings (hr)
LOW HAZARD AREAS A storage area not greater than 450m^2 Workshops and reprographic rooms Kitchens Transformer, switchgear and battery rooms for low, or extra low voltage equipment	½
MEDIUM HAZARD AREAS Loading bay A storage area greater than 450m^2 Room housing a service not listed as low or high hazard Car park not greater than 450m^2 within or adjoining an office building	1
HIGH HAZARD AREAS Car park greater than 450m^2 within or adjoining an office building Boiler and fuel storage room Transformer and switchgear room for normal voltage equipment Room for internal combustion engine Any other high hazard area	1

NOTE If the period of fire resistance recommended in Table 4.7 is lower than that given in Approved Document B2/3/4 for structural members generally, the latter should be used.

visited or because of the consequences of a malfunction of the equipment housed.

In an area of higher fire risk, it is important that that part of the travel distance within the room be strictly limited. With only one exit from the room, there should be no more than 6m between the door and the point furthest from it. If there is more than one exit, 12m is the recommended maximum distance from an exit.

Any room in which flammable liquids or gases are stored or used should have a cill at the doorway and an interceptor for any floor drainage. All service accommodation should have adequate ventilation to the external air. Any exit from service accommodation which is situated at basement level should not be sited so as to prejudice escape through any final exit from the building.

High fire-risk areas should have a protected lobby or corridor between them and any protected stairway forming part of an escape route.

Transformer and Switchgear Rooms.

Unless situated on the roof or away from the main building, these rooms are best sited on an external wall and entered from outside the building. There should be adequate ventilation and the travel distance within the room limited to 6m if there is one exit, otherwise 12m.

Protected Circuits

These should be provided for all equipment which contributes to safety in the event of fire, notably, alarms, detectors and escape route lighting. The circuits should consist of mineral insulated metal sheathed or other heat-resisting cable. The route chosen for the wiring should only pass through areas with a negligible fire risk and the circuits should be independent of any other wiring.

Escape Route Lighting

Office occupants will be familiar with the normal circulation routes of the building; fire drills should have given them a sound knowledge of the escape routes. Nevertheless, there may be visitors in the building and employees can become disorientated if a real emergency occurs in darkness.

It is therefore necessary to ensure that all parts of the escape route will be adequately lit day and night. The lighting may be permanently on or may be automatically switched should the mains fail. It should either be supplied by an independent circuit which cannot be affected by power failure either of the mains or due to fire in the building. This may necessitate the provision of central batteries or a stand-by generator system. The Code recommends that emergency lighting systems should conform to BS 5266: Pt 1: "Code of Practice for emergency lighting".

The Code recommends that escape lighting should always be provided within the following areas:

- Underground and windowless accommodation
- Internal corridors which exceed 30m in length and have no borrowed light.
- Stairways in buildings higher than 18m and all stairways in a central core.

Boiler Rooms and Fuel Storage

In new buildings, boilers will normally be installed in accordance with appropriate British Standards depending on the fuel burned. Boiler installations not covered by British Standards should have adequate provision for smoke venting and in both boiler rooms and fuel storage rooms travel distances should not be greater than 6m if there is only one exit or 12m with two or more.

There are British Standards (BS 5410: Pts 1 & 2 and BS 799: Pt 5) which give guidance on the safe storage of oil fuel. Solid fuel bunker walls should be sufficiently thick to prevent the fuel being heated by adjacent boilers or steam pipes.

If liquefied petroleum is to be stored in bulk the storage space should be constructed in accordance with advice in note CS 5, "The storage of liquefied petroleum gas at fixed installations", issued by the Health and Safety Executive.

Ventilation Systems
Mechanical ventilation and air conditioning systems should be installed in accordance with the BS CP 413: 1973. The air distribution system should be arranged so that the normal airflow pattern is away from protected escape routes and exits. If the system recirculates air, it should be provided with smoke detectors in the extract ducts which, when smoke is present, will direct extracted air to the outside of the building and stop recirculation. The system should also be compatible with any escape route pressurisation system fitted.

Lifts
Lifts are not to be considered part of the means of escape, except that, in the event of fire, one or more may be designated for use by the chairborne disabled. Fire-fighting lifts are outside the scope of the Mandatory Rules.

Lift motor rooms are better located at the top of the lift shaft; if the motor room must be at the foot of the shaft, fire separation between it and the shaft is necessary. The lift shaft will have to be constructed as a protected shaft (see Approved Document B2/3/4). Lift motor rooms should comply with BS 5655.

Waste Disposal
In office buildings the main waste material may well be paper and chutes may be installed to deal with this and any other kind of waste. These are "protected shafts" in the terms of Approved Document B2/3/4 and reference should be made to paras. 6.29 to 6.34 as well as to Appendix E of that document.

Requirement H4 of Schedule 1 of the regulations requires the provision of satisfactory means of storing solid waste in a building at or near the base of the chute; there must also be adequate access to the storage chamber from the street for the removal of waste. Recommendations for satisfying these requirements will be found in Approved Document H4.

The Approved Document and the Code both cite BS 5906: 1980 which is the Code of Practice for storage and on site treatment of solid waste from buildings.

Incinerators
Two types of incinerator may be installed in an office building; there may be sanitary incinerators in toilet accommodation or an incinerator for disposing of combustible bulk waste. Either type may be fired by gas or electricity.

As both types of incinerator are "heat-producing appliances" the installation and associated flues are subject to requirements J2 and J3 of Schedule 1 of the regulations. General recommendations on how to comply are to be found in Approved Document J1/2/3 although this document does not specifically mention incinerators. The only recommendation of significance in the Code is that if capacity exceeds $0.08m^3$ it is preferable for the incinerator to be located in a separate building.

Storage Facilities
Office buildings frequently are provided with areas where furniture, stationery and other items of office equipment can be loaded and unloaded and stored. There may be associated sorting and packing rooms. Special precautions will be needed should any of the materials regularly being received or dispatched be of a highly flammable nature. Otherwise, only when the storage area is larger than $450m^2$ or is situated below ground level does the Code recommend provision for smoke venting.

A storage area is better situated adjacent to an external wall and in a location where smoke discharging from it cannot jeopardise an escape route from the building. There should be adequate fire resistance in the structure separating these areas from the remainder of the building; the recommended period of fire resistance will be higher if the goods stored or handled are of a hazardous nature.

Car Parks

There are requirements under the Petroleum (Consolidation) Act 1928 controlling various aspects of the design of large car parks (greater than $450m^2$ in area) which adjoin or are within a building. Recommendations for multi-storey car parks will be found in Approved Document B2/3/4, paras. 6.61 and 6.62. The Code refers to an HMSO publication dated 1968 and entitled *Model Code of Principles of Construction and Licensing Conditions for the Storage of Cans, Drums, and Other Receptacles; Petrol Filling Stations, Basement and Sub-Basement Garages and Underground Car Parks*.

The Code recommends that any car park within or adjoining an office building, irrespective of any requirement for it to be licensed or not, should be constructed in accordance with the following recommendations:

- Any permitted access between the garage or car park and the office accommodation should incorporate a protected lobby.
- Any opening provided for lighting or ventilating the car park should be so situated that it does not endanger any escape route or the final exit for any part of the office.
- There should be adequate provision for venting smoke.

Workshops and Reprographic Rooms

Some large offices may have workshops for the repair of furniture or the maintenance of office equipment. There may also be reprographic, photographic or small printing facilities; flammable materials may be used or stored in these areas. Only when these facilities are extensive does the Code recommend that they are located in rooms which are separated from the offices.

If highly flammable materials are used or stored, the area will constitute an "area of high fire risk" and special requirements may be imposed by the fire authority or by inspectors of the Health and Safety Executive.

Kitchens, Staff Restaurants and Canteens

The main fire risk is in the kitchen which, if possible, should be separated from the restaurant or canteen by fire-resisting construction (half-hour period). Where this is impossible because of, for example, a serving counter, the fire protection should be around both areas and the means of escape should be directly from the restaurant or canteen.

The kitchen should have at least one escape route which is independent of a service door to the restaurant or canteen. No escape route from the restaurant or canteen should pass through the kitchen.

THE REMAINDER OF THE CODE

It should be noted that the Code contains another section, called "Fire protection facilities" which is not subject to the Mandatory Rules for Means of Escape. Nevertheless, it will prove to be important to designers from the points of view of fire certification and building insurance.

It deals with fire alarm and detection systems, fire extinguishing systems such as sprinklers and fire mains, methods of smoke control and the provision of fire lifts and firemen's lobbies. In section 30, "Smoke control provisions", advice will be found on the ventilation of protected lobbies and basements. There is also a reference to BS 5588: Pt 4 in a short section dealing with smoke control by pressurisation; this form of smoke control is mentioned in one or two places in clauses which are subject to the Mandatory Rules.

APPENDIX
Relevant clauses of BS 5588: Part 3: 1983

The following clauses of BS 5588: Part 3: 1983 are cited in para. 4.3 of Section 4 of the Mandatory Rules. They are listed here together with an indication of the subject dealt with by each clause.

SECTION 1 General

- 2 Definitions

SECTION 2 Analysis of the problem

- 4 Planning in relation to fire
  - 1 Spread of fire
  - 2 Smoke – effect on people
  - 3 Site planning
  - 4 Mixed user developments
  - 5 Offices in different occupation
  - 6 Internal subdivision
    - compartments
    - open vertical planning
    - open plan offices (horizontal)
    - cellular with corridors
  - 7 Ancillary and high risk areas
- 5 Escape from fire
  - 1 General
  - 2 Evacuation (full or partial)
  - 3 Disabled persons

SECTION 3 Planning escape from office areas

- 6 General
- 7 Escape routes within a storey and exits from a storey
  - 1 Travel distance and number of escape routes
  - 2 Number of exits from a storey or room
  - 3 Width of exits and escape routes
  - 4 Corridors
  - 5 Offices in different occupation
  - 6 Escape across a flat roof
- 8 Stairways and final exits
  - 1 Accommodation stairways and escalators
  - 2 Number of protected stairways
  - 3 Siting
  - 4 Width
  - 5 Enclosure
  - 6 Stairways to basements
  - 7 Access lobbies and protected corridors
  - 8 External escape stairways
  - 9 Discharge from stairways and final exits
  - 10 Examples

CHAPTER 3

Special Provisions for Disabled People

INTRODUCTION

Part M of the Schedule to the Building Regulations requires that "reasonable provision shall be made to enable disabled people to gain access to the relevant premises, and to those parts of the relevant premises to which it is reasonable to provide access." This applies to buildings being altered or extended as well as to new construction.

Relevant premises and parts include all storeys of an office and a shop and the principal entrance storey of a factory, educational building and a public building. Where it is a requirement that access be provided, it is necessary also to provide adequate means of escape. Current thinking goes further than Building Regulations in believing that all buildings should be reasonably accessible to disabled people.

For the purposes of Part M, "a disabled person" is one who has a physical impairment which limits ability to walk or necessitates the use of a wheelchair. However, adequate means of escape should be provided for all potential occupants of buildings which are subject to requirement B1 of Building Regulations or to the fire certification requirements of the Fire Precautions Act, as amended. This implies that means of escape provisions should take into account the needs of those disabled by subnormal vision (including blind people who cannot read fire safety signs and notices) and those whose hearing is impaired (who may not hear a fire alarm or verbal instructions to escape given by staff or over a public address system).

The British Standard BS 5588: Part 8: 1988 is the Code of Practice for means of escape for disabled people. The recommendations are supplementary to those in Parts 2 and 3 of BS 5588 but do not apply to dwellings. (See BS 5619: 1978 the Code of Practice for the design of housing for the convenience of disabled people.) Buildings which are designed specifically for the disabled will generally be given more extensive provisions than are recommended in Part 8. The main recommendations of the BS which are "structural" are summarised here but it should be remembered that the safety of all building occupants, and of the disabled in particular, depends greatly on good management practices and the proper training of staff.

DEFINITIONS

In addition to terms already defined in the fire precaution Codes for different classes of building, the following should be noted.

EVACUATION LIFT
A lift which is enclosed in a shaft having fire resistance equivalent to that of a protected stairway and with special features which enable it to be used safely for the evacuation of disabled people in a fire. The evacuation lift may be a fire-fighting lift (q.v.).

FIRE-FIGHTING LIFT
A lift within a protected shaft and with special controls which enable it to be taken over by the fire brigade and used to gain access for fire-fighting.

REFUGE
A protected zone on an escape route with space for a wheelchair and occupant to wait safely for assistance. The space will be so located that it does not impede the general escape flow along the route and will not be rendered unusable by a build-up of smoke.

WHEELCHAIR STAIRLIFT
An inclined track, associated with a stairway, with a powered carriage for transporting a wheelchair and its occupant upwards or downwards between levels of a building.

RECOMMENDATIONS

Horizontal escape routes
Along an escape route, changes in floor level within a storey should be avoided. If differences in level do occur, ramps rather than steps should be used and a warning of the change given by means of signs and in some circumstances changes in the texture of finishes. See also Part K1 of the Regulations and BS 5810: 1979.

Disabled people may have to wait for help at certain points along the escape route. Negotiating a stairway, for example, would be impossible for someone in a wheelchair. Refuges should therefore be provided within which it will be quite safe to wait for a limited period. If the storey is divided into compartments, a compartment which is not the one in which the fire has occurred may be regarded as a refuge. More usually, refuges will be within protected areas – a stairway enclosure, a lobby approach to a stairway or a protected corridor.

Except in single-storey buildings and certain small buildings defined in the next paragraph, there should be a refuge associated with every protected stairway on every storey to which the disabled have access. This requirement does not apply to a storey all exits from which are final exits. Refuge space should also be provided within the protective enclosure of an evacuation lift adjacent to the lift doors.

The small buildings in which refuges are not necessary are those which comprise not more than a ground and first storey. (The ground storey may have a basement below it.) None of the storeys should have a floor area exceeding $280m^2$ and the entire building should be in a single occupancy.

The Code recommends a minimum area of 700 mm × 1200 mm plus space for manoeuvring; it suggests the ideal is 900 mm × 1400 mm. The Code visualises only one such area on each storey per stairway or per evacuation lift. In some circumstances, however, this may not be deemed sufficient. The refuge area should be clear of the flow path of the means of escape.

Vertical escape routes
For disabled people the vertical escape route will comprise either a protected stairway, a ramp within a protected enclosure or an evacuation lift. All stairways and ramps in new buildings must be constructed to comply with requirement K1 of Building Regulations; AD(K) gives advice on satisfying K1. AD(M) Appendix A also includes advice about the design of ramps suitable for wheelchair users.

STAIRWAYS
In addition to the recommendations of AD(K1), handrails on stairways to be used by disabled persons should be continuous over the length of a flight; they should be mounted sufficiently clear of the wall, and have a convenient shape in cross-section, so that they can be easily and confidently gripped by the user.

The Code recommends the use of contrasting colours for wall and handrail and for stair nosings to help the visually handicapped to see them. It is also suggested that floor finishes to the thresholds at the top and bottom of each flight be textured to provide a tactile warning of the change of level.

The storey on which the final exit is located should have this fact indicated by having the word 'EXIT' clearly displayed on or near the handrail. On all other storeys arrows, similarly positioned, should point upwards or downwards to indicate direction to the final exit storey.

If the stairway is equipped with a stairlift for wheelchairs, no part of the installation should encroach on the means of escape. Such stairlifts should not be used by the wheelchair occupants as a means of escape.

Ramps

Wheelchair users and the ambulant disabled can negotiate a suitable ramp more easily than a stairway. The Appendix to AD(M) contains guidance, but generally a gradient not steeper than 1:12 will be suitable; the ramp should be surfaced with a non-slip material.

Handrails similar to those recommended for stairways should be provided as should the arrow and exit signs and the tactile thresholds warning of change of slope.

Lifts

Evacuation lifts are not a requirement for the disabled but one or more may be provided; if it is intended (or there is a requirement) to provide a fire-fighting lift, this can also serve as an evacuation lift. The Secretary of State, in Determinations concerned with the provision of evacuation lifts, has made it clear that, in normal circumstances, it is reasonable to expect help to be available so that wheelchair occupants can be helped down a stairway.

Lifts provided for evacuation or fire-fighting must be so installed and protected that they will continue to operate for a significant time after fire has broken out in a building. British Standards BS 5810 and BS 5655: Parts 1 & 2 make recommendations.

The lift should be in a protected enclosure which should include a lift lobby at each storey level constructed as a protected lobby. The standard of protection should be the same as is required for a protected stairway according to the use and size of the building. No door giving access to the lift lobby from a storey should be in the wall facing the lift door(s). The lobby should be large enough to provide space for use as a refuge by a wheelchair occupant waiting for assistance. The route leading from the lift to the final exit from the building should be protected.

The power supply to the lift motor should be independent of the main supply to the building. It should run within the protected enclosure or be in protected cable. If it cannot be completely isolated from the power supply to the remainder of the building, a standby generator may be needed which will start automatically in the event of power failure. Hydraulic lifts, serving only a ground and first storey, need not have this alternative power supply.

Except when a lift serves two storeys only, the switching arrangement must allow authorised persons to take over control. This is achieved by a switch beside the lift door on the final exit storey, clearly marked "Evacuation Lift", which, when pressed, makes the call buttons on all other storeys inoperative and returns the lift to the exit storey. The lift car control panel is then used to operate the lift. It will be necessary to have an independent communication system which allows a disabled person on any storey to call for the lift. An authorised person is either a fireman or a staff member delegated and trained to this duty by the building management.

Index